Life in the Universe

Springer
Berlin
Heidelberg
New York
Hong Kong
London
Milan
Paris
Tokyo

Advances in Astrobiology and Biogeophysics

springeronline.com

This series aims to report new developments in research and teaching in the interdisciplinary fields of astrobiology and biogeophysics. This encompasses all aspects of research into the origins of life – from the creation of matter to the emergence of complex life forms – and the study of both structure and evolution of planetary ecosystems under a given set of astro- and geophysical parameters. The methods considered can be of theoretical, computational, experimental and observational nature. Preference will be given to proposals where the manuscript puts particular emphasis on the overall readability in view of the broad spectrum of scientific backgrounds involved in astrobiology and biogeophysics.
The type of material considered for publication includes:

- Topical monographs
- Lectures on a new field, or presenting a new angle on a classical field
- Suitably edited research reports
- Compilations of selected papers from meetings that are devoted to specific topics

The timeliness of a manuscript is more important than its form which may be unfinished or tentative. Publication in this new series is thus intended as a service to the international scientific community in that the publisher, Springer-Verlag, offers global promotion and distribution of documents which otherwise have a restricted readership. Once published and copyrighted, they can be documented in the scientific literature.

Dirk Schulze-Makuch Louis N. Irwin

Life in the Universe

Expectations and Constraints

Springer

Professor Dirk Schulze-Makuch
Professor Louis N. Irwin
University of Texas at El Paso
500 W University Avenue
El Paso, Texas 79968-0555, USA

Cover picture by Chris D'Arcy, Dragon Wine Illustrations, El Paso, Texas

Cataloging-in-Publication Data:
Schulze-Makuch, Dirk, 1964-, Life in the universe: expectation and constraints/ Dirk Schulze-Makuch, Louis N. Irwin. p.cm. – (advances in astrobiology and biogeophysics) Includes bibliographical references (p.). ISBN 3-540-20627-2 (alk. paper) 1. Life (Biology) 2. Life on other planets I. Irwin, Louis N. II. Title. IIII. Series. QH341.S339 2004 576.8'39–dc22

Bibliographic information published by Die Deutsche Bibliothek Die Deutsche Bibliothek lists this publication in the Deutsche Nationalbibliografie; detailed bibliographic data is available in the Internet at <http://dnb.ddb.de>

ISSN 1610-8957

ISBN 3-540-20627-2 Springer-Verlag Berlin Heidelberg New York

Springer-Verlag is a part of Springer Science+Business Media
springeronline.com

Printed in Germany

Cover design: Erich Kirchner, Heidelberg

Printed on acid-free paper 54/3141/ts - 5 4 3 2 1 0

Preface

Exobiology, or life on other worlds, has been a source of speculation since ancient times. In recent years, the National Aeronautics and Space Administration (NASA), as a matter of public policy, has elevated the loose-knit collection of models, simulations, experiments, and speculations relevant to the origin, distribution, evolution, and fate of life in the universe to a formal science, which it denotes as astrobiology. Exobiology and its derivative form, astrobiology, now form a significant part of the scientific program for all the space exploring nations. Russia and the United States, the pioneers in space, have been joined by the European Space Agency, both independently and collaboratively, with a host of new missions. In Asia, the Japanese have entered the exploration of space, as have the Chinese with their first astronaut in space. The International Space Station and the Cassini–Huygens Mission to Saturn testify to the vigor and maturity of the international effort in space, much of which is geared toward the study of life in space and on other worlds.

With this upsurge of interest has come a host of books, mostly for a broad audience, and often for popular consumption. Nearly all have been enthusiastic about the possibility of life on other worlds. But scientific depth has often been sacrificed in favor of the laudable goal of engaging a broad audience.

This book embraces the conviction that life is not restricted to our planet. We accept as plausible that, with 10^{11} stars in our galaxy, and each frequency component of the Drake Equation conservatively set to 0.01, there may be 10^5 abodes for life in our galaxy alone. And there are billions of galaxies. While computations such as these are debatable (if unresolvable at the present time), we do not take issue with their general conclusion. Rather, the objective of this book is to analyze in critical scientific detail the fundamental, commonly-held assumptions about life beyond Earth – particularly those relating to the probable cosmic preference for carbon-based life, the overwhelming focus on water as the preferred solvent for life, and the relative merits of different forms of energy for the sustenance of life. While we do assume that extraterrestrial life exists, we take nothing about its nature for granted.

The need for these three pillars of life – carbon, water, and energy – to be analyzed with rigor is particularly critical at this time, as the Galileo Mission has just ended with an unprecedented storehouse of data on the Jovian system, with the Cassini–Huygens Mission on the brink of an even greater stream of data from the Saturnian System, and as the Mars Odyssey Orbiter and new landers currently glean fascinating new information from that planet on a daily basis. Missions to Mars are in the pipeline every two years for the foreseeable future; another mission to the icy moons of Jupiter, and probes to Pluto, are on the drawing board; and sample return missions to Venus are even being discussed. These billion dollar projects will all rely on remote detection technologies. It is critically important that we search intelligently and comprehensively for the appropriate biosignatures and geoindicators that could mark or suggest the presence of life. This book is written with a particular emphasis on the scientific rationale for what we should be seeking, and how we should be looking for it.

We have also tried to write a book that treats biology with greater sophistication than most treatises on the subject. Certain assumptions commonly recur in astrobiological literature: life originates quickly once it has the opportunity to do so, but takes a long time to evolve to multicellularity; most extraterrestrial life is probably microbial; darwinian evolution is a requirement for life; the evolution of intelligence is an improbable event. Some of these assertions are valid; others have little foundation. We have drawn on evolutionary and ecological theory to critically evaluate these and related issues, and have documented our arguments so that the skeptical reader can pursue them to their source.

This is not a comprehensive overview of astrobiology. We have not dealt at all with the origin of life, as many excellent books are available on that subject. Nor have we written about the future of life, since searching for the unseen life that currently exists is challenging enough for now. While we have endeavored to carefully document all our statements of fact, and all the speculations that are not original to us, our bibliography is selective. Thousands of talented scientists and visionary science writers for the general public are publishing in this field, and we regret with apology that we have not been able to cite all the worthy authors among them.

This book had its origin in discussions with our Astrobiology Graduate Seminar of 2001, and was tested in draft form with our seminar in 2003. We thank the students in those classes for their stimulating input. We also thank Philip Goodell, Ousama Abbas, and Keith Pannell of the University of Texas at El Paso for reviewing selected parts of the text. Two anonymous reviewers and Kevin Plaxco, at the University of California at Santa Barbara reviewed the complete text and made many valuable suggestions. Our editor at Springer-Verlag, Christian Caron, provided excellent critical oversight and support. The final result is, of course, solely our responsibility. Finally, we thank NASA and the J. Edward and Helen M.C. Stern Endowment for financially supporting our activities when we first became involved in the field of astrobiology.

Dirk Schulze-Makuch would like to thank his wife, Joanna, and his children Nikolas, Alexander, and Alicia for their patience and understanding. He also thanks his students Huade Guan, Wayne Belzer, Benjamin Diaz, and Tanya Lehner for fruitful discussions and a willingness to bounce around ideas, and Sigung Hasting Albo for teaching him the concept of Qi.

Louis Irwin thanks his wife, Carol, for her erroneous concern that he works too hard, and all his graduate students who have been neglected during the writing of this work.

> "Imagination is more important than knowledge. Knowledge is limited. Imagination encircles the world."
>
> Albert Einstein

Dedication

To an embryonic individual of a species on a water-rich planet in an otherwise unremarkable solar system at an outlying part of the galaxy in an unexceptional part of the universe. The ultrasound image below shows one incipient being (the unborn daughter or son of the first author) with a length of 0.58 cm, less than one hundredth of its adult size and still in the body of its mother, but already with a beating heart (area between cross-hairs) – a remarkable consequence of about 4 billion years of evolution leading to a species just beginning to explore its universe – and with every new birth the hope that he or she will bring us closer to an understanding of the world in which we live.

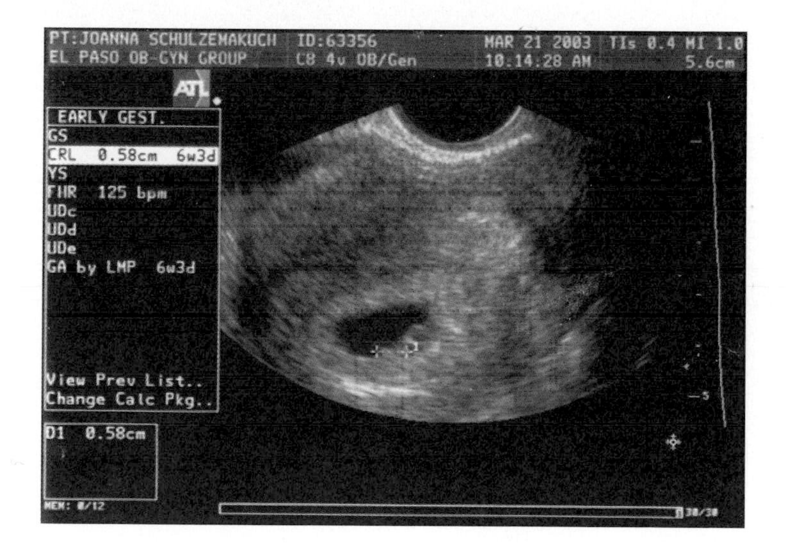

Contents

1 Introduction

Astrobiology studies the origin, evolution, distribution, and fate of life throughout the universe, with no direct evidence that life exists anywhere in the universe other than on Earth. But there are compelling reasons to assume that life exists pervasively throughout the cosmos. That assumption is derived from empirical observations on the nature of the universe and the natural laws that govern it, from analysis of the history and properties of the one case of life that we do know, and on a logical integration of fact and theory. The science of astrobiology is thus as strong, if not as revolutionary, as Darwin's theory of evolution before fossil humans were found to prove our animal origins; as firm, if not as precise, as the astronomical predictions that Neptune must exist before it was detected; and, in our view, as certain as the conclusion that the world was a sphere before Magellan sailed around it.

This book sets forth the argument that life occurs redundantly throughout the universe. It further makes predictions about some likely characteristics of that life in most cases, explores the limits of diversity that might be found in forms of life on other worlds, and attempts to strain conventional thinking about the fundamental nature of living systems. At the same time, this book asks for no suspension of belief in or extension beyond the laws of chemistry and physics as we understand them now. It does not make predictions of a specific nature where no basis for specificity exists. We offer our assessment about probabilities, but base those assessments on facts open to verification and a line of reasoning that invites the critical assessment of our fellow scientists. Like Darwin's arguments about the mechanism of evolution, we know that our vision of life in the universe will change through subsequent insights and observations. As the predictable discovery of Neptune gave no indication of the altogether unpredictable planet Pluto yet to be discovered, we realize that surprises not anticipated by us will emerge when the reality of life on other worlds is confirmed. And finally, like Magellan, we fully expect some of our calculations about life in the universe to miss their mark. But we do believe we have sketched a vision of cosmic biology that is tenable and therefore of predictive value in designing missions to search for and detect the life that is surely out there.

The argument that life exists on other worlds is straightforward and simple. It begins with the definition of life as a self-perpetuating organization of complex chemistry that uses free energy to maintain disequilibrium with its environment. It continues with the observation that wherever chemical heterogeneity and a source of free energy are found, the capability for life exists. It notes that our own planet, which had an abundance of both energy and complex chemistry from its earliest age, gave rise to life (or was able to sustain life introduced from another place) almost as soon as the heavy bombardment of the planet receded. It assumes that the laws of chemistry and physics act in the same way throughout the universe. It points to the vast numbers of stars in the universe, and the possibility that the total number of planets will be even larger. The argument concludes: even if the probability on any given planetary body is low that an appropriate combination of energy and chemistry is available to enable the development of sufficient complexity for life to emerge, the enormous number of planetary bodies that must exist in the visible part of the universe alone strongly suggests that life has arisen redundantly. Since the physical laws of nature pertain equally, we assume, over the entire extent of the universe, wherever those laws allow the formation of life to occur, it will. Hence, life must be widespread as well as high in numerical frequency.

We should emphasize that we do not argue that life is common. The complexities of form and function that constitute the living state are highly improbable in a statistical sense, and probably arise only under a restricted set of circumstances. The second purpose of this treatise is to critically examine what those circumstances are. To the extent that our solar system is exemplary (we cannot yet say that it is typical), the conditions that exist on Earth, where a large range of microscopic to macroscopic forms of life have diversified, appear to be very rare. We think, therefore, that the extent of biodiversity that we see on Earth is very seldom seen anywhere else. Among the organisms that thrive on our planet, however, are many microscopic forms that potentially could occupy a number of other sites in our solar system, as carbon polymer and water-based life essentially as we know it. In addition, however, there are circumstances substantially unlike those with which we are familiar on Earth, under which life in forms unknown to us could arise and exist, in theory. Those circumstances are found within our solar system, and are likely to be found beyond it in abundance. The circumstances that would allow for the origin and persistence of life are not unlimited, however. Much of this work is devoted to assessing what those limits might be.

Our analysis focuses primarily on four facets that are essential to life: energy, chemistry, solvent, and habitat. To provide the reader with an overview at the outset, a brief abstract of our analysis of the possibilities and limitations of each of these facets is given below.

Energy in many forms is abundant throughout the universe. Electromagnetic energy at wavelengths visible to humans is a prominent product of the fusion reaction in all the visible stars. On Earth, a photosynthetic mechanism has evolved to capture that energy and transform it into chemical bonds with an efficiency that is difficult for any other form of energy to match. Where light is available, it thus provides an efficient, isothermal source of energy well matched to the needs of living systems. However, both inorganic and organic chemical bonds contain energy that is harvested by all non-photoautotrophs on Earth, so far as we are aware. As long as these sources of chemical energy remain available, either from cycling or a reserve not yet exhausted, they likewise provide an efficient basis for bioenergetics. Other forms of energy could in principle be substituted to varying degrees for the light and chemical energy that support the forms of life with which we are familiar. Our theoretical calculations suggest that osmotic and ionic gradients, and the kinetic motion of convection currents, provide plausible alternatives. Thermal gradients are among the most widely available sources of energy flow, but the gradients are easily degraded and are thermodynamically inefficient. Magnetospheric energy, gravity, pressure, and other exotic forms of energy likewise could conceivably be harvested by living systems, but the amount of energy that they provide within our solar system generally does not appear to make them competitive with light, chemistry, osmotic and ionic gradients, or convective currents as likely sources of free energy for the support of living systems.

All life as we know it resides in complex polymeric chemistry based on a covalently bonded carbon backbone. A systematic examination of carbon chemistry provides an impressive list of advantages that carbon has over any other compound, not only in forming the vast array of molecules required for complex systems, but by enabling the right combination of stability and flexibility for molecular transformations that underlie the dynamic complexity of life. In aqueous systems at temperatures common on Earth, carbon is so far superior to any other atom as a polymeric unit, that it has come to be the only basis for the structure of biomolecules essential for all basic metabolic processes. Silicon is the one other atom with properties similar to carbon, and its potential usefulness in living systems is shown by the fact that it too is an important constituent of many living cells. In most cases, it serves a rather passive structural role, as in the cell walls of plants, and the exoskeleton of diatoms and some other organisms. These examples could represent residual functions from a time in the history of life when silicon played a more central role, only to be replaced more effectively by carbon at a later stage. A detailed look at the chemistry of polymeric silicon reveals that it conceivably could have the combination of stability and lability exhibited by carbon, but under very different conditions, both at temperatures much higher and much lower, and in the presence of solvents other than water. Carbon bonds with oxygen and nitrogen to form parts of the polymeric chains of biomolecules, and mixed atomic backbones involving other compounds are a possibility. They already occur in some biomolecules of terrestrial life such as DNA but may be much more common elsewhere. A few other atoms have the capacity for the formation of covalent polymers, but they either occur in such low abundance, or have such inferior characteristics, that they seem a highly unlikely basis for an alternative living system.

Life as we know it requires a liquid medium. It can survive periods of dehydration, but appears to need a liquid for its dynamic transactions. We examine in some detail why life is much less likely to reside exclusively in a gaseous or solid medium. We also consider whether water is the only suitable solvent for a living system. Water does have some striking advantages, particularly with respect to carbon-based molecular interactions. At temperatures and pressures prevailing on Earth, and beneath the surfaces of numerous other planets and planetoids in our solar system, water can exist in liquid form, and thereby provide the potential reservoir that carbon-based molecules need for their vast array of interactions. On the other hand, most of the solar system, like most of the universe, is very unlike Earth. For smaller planetary bodies distant from a star, temperatures are much colder than on Earth. This probably represents the vast majority of planetary bodies. At those sites, water cannot be liquid (absent a source of internal heat), but methane, ammonia, ethane, methyl alcohol, and related organic compounds might be. In principle, many of them are compatible with carbon-based polymeric chemistry, and thus should be considered as possible solvents capable of supporting life. On very large planetary bodies, or on those that are tectonically active or close to a central star, very high temperatures may prevail. Under those circumstances other compounds can exist in the liquid form. In some of those cases, silicon-based polymers appear more feasible. There is no question that water is an excellent solvent for living systems, but under conditions where it cannot exist as a liquid, a few other solvents can exist in that state, and could support living processes.

Habitats can be divided grossly into those that are constant and those that are variable. The surface of a planetary body under rare circumstances as on Earth may be quite variable, providing the opportunity for fragmentation of the environment into a great variety of subhabitats with specific but periodically changing characteristics. These variations and their changes over time represent selective pressures that generate, through the evolutionary process, a great variety of living forms. When, as on Earth, energy and appropriate chemical environments are abundant, life can assume macrobiotic forms of great complexity. The cost of this biodiversity and complexity, however, is frequent extinction, as changing conditions in variable habitats often render biological features that were advantageous under one set of circumstances, suddenly disadvantageous under others. The cycle of speciation followed by extinction generates the biodiversity and great deviation from primordial forms of life with which we are familiar. We must remind ourselves, however, that the primordial forms are still with us as well. They are sequestered primarily below the surface, where the constancy of conditions places a premium on stabilizing selection, or the retention of successful living processes that have experienced no pressure for change for a long time. Only now are we beginning to appreciate the vastness of this subterranean, unseen biosphere; but it probably represents the most favorable and most common habitat for life throughout the universe as a whole. The consequences of subsurface life are two-fold: First, the minute size of the living spaces available restrict the size of living organisms to microscopic dimensions. Secondly, the long-term stability of the environment places a premium on stabilizing selection, which probably maintains life in an ancestral form. In those rare planetary bodies that have gaseous atmospheres, life may exist as well, but it probably is microscopic in that sphere also, though is much more likely to have deviated significantly from its ancestral form.

In the chapters that follow, we elaborate on these arguments in greater detail and discuss how life can be detected. Our vision of astrobiology is driven by our sense that life, like all of nature, is knowable in principle wherever it exists. We are strongly persuaded by scientific evidence and logic that it exists in profusion on other worlds. We believe it is likely that it exists elsewhere in our solar system in at least a few instances, though probably in microbial form. We hope to see the day when this belief is confirmed by direct evidence. If we do not, we nonetheless are confident that a perceptive form of life somewhere, someday, will encounter life on a world other than its own. How similar or how different those forms of life will be is one of the most enticing questions of our age. This book is meant to explore the range of answers that might be offered.

2 Definition of Life

The definition of life is a long-standing debate with no general scientific consensus to be expected any time soon. The underlying problem is that living systems use compounds that are abundant in the surrounding environment and processes that are not intrinsically different from reactions that occur abiologically. There does not appear to exist a single characteristic property that is both intrinsic and unique to life. Rather we have to argue that life meets certain standards, or that it qualifies by the collective presence of a certain set of characteristics. The threshold for meeting this standard sounds arbitrary, and may well be arbitrary in the sense that life presumably arose through a long sequence of "emergent events", each at a greater level of molecular complexity and order (Hazen 2002). If that notion is correct, any rigid distinction between life and non-life is a matter of subjective judgment. While our everyday experience with life on Earth makes the distinction between the living and non-living for the most part unambiguous, a consideration of life on other worlds, where conditions may be different, and/or where life may have evolved from its inorganic precedents to a lesser degree, requires us to formulate a more formal and objective definition for life. Before doing so, we will first address the limitations of commonplace assumptions about what constitutes life.

2.1 Problems with Common Assumptions about the Nature of Life

Dictionary definitions of life (e.g. Random House 1987) typically refer to its characteristics and processes, such as metabolism, growth, reproduction, and adaptation to the environment. This form of definition is generally followed by some biology textbooks (Campbell 1996, Raven and Johnson 1999), while others – tacitly admitting the difficulty of defining life – refer instead to its "unifying principles" (Curtis and Barnes 1989) or its "emergent properties" (Purves et al. 1998). Because a process definition is the most traditional and intuitive way to look at life, we will first discuss each of the processes traditionally associated with life, and point out similarities to characteristics and processes in non-biological systems. We will then consider attempts to circumvent the difficulties of the traditional definitions, and finally offer our own definition as one with greater application to the issues with which astrobiology must deal, such as remote detection and alternative forms of life on other worlds.

The consumption or transformation of energy is a central point in all traditional definitions of life. Energy metabolism in its most basic form consists of a collection of chemical reactions that yield energy by electron transfer. Living organisms obtain energy from light by photosynthesis or by other electron transfer reactions associated with chemoautotrophy (extraction of energy from non-biological molecules) or heterotrophy (extraction of energy from molecules synthesized by other living organisms). Energy is then stored in the phosphate bonds of ATP, GTP, creatine phosphate, or similar molecules until it is used for various purposes. However, inorganic analogs of these processes are well known. Electrons can be lifted into higher energy levels by various forms of energy, such as heat or ultraviolet radiation. When the electrons fall back to their lower energy levels, the energy difference between these levels is released. When ions absorb energy and release it again in the form of light, this is known as luminescence. If the absorbed energy is "frozen in", and released only upon heating, the process is known as thermoluminescence. Phosphorescence and fluorescence are special cases of luminescence and describe the phenomenon of continued emission of light after irradiation is terminated. The excited-state lifetime of fluorescence is about 10^{-8} s, while the lifetime of phosphorescence is about 10^{-3}–10^{3} s. The wavelength of the luminescent light is longer than the wavelength of the exciting radiation, thus representing transformation. The phenomenon of luminescence is linked to many factors such as lattice defects or the presence of foreign ions that often substitute for major elements in the crystal structure and function as activators (Mason and Berry 1968). Common minerals with the property of luminescence include gypsum and calcite. Another possibility for storing energy in the form of heat is seen in clay minerals with interlayer sites. The interlayer water and OH-groups are suitable for storing heat energy due to their high heat capacity. Thus, nonliving substances can transfer external energy into energy-yielding transitions that under some circumstances can be maintained as potential energy, just as living organisms do (Schulze-Makuch et al. 2002).

Another traditionally regarded property of life is growth. But just as cells grow in favorable environments with nutrients available, inorganic crystals can grow so long as ion sources and favorable surroundings are provided. Furthermore, just as the development of living organisms follows a regulated trajectory, so the process of local surface reversibility regulates the course of silicate or metal oxide crystals that grow in aqueous solutions (Cairns-Smith 1982).

The third traditionally defined property of life is reproduction, which entails both multiplication of form and transmission of information. The visible consequence of reproduction in living organisms is the multiplication of individuals into offspring of like form and function. Mineral crystals do not reproduce in a biological sense, but when they reach a certain size they break apart along their cleavage planes. This is clearly a form of multiplication. The consequence of biological reproduction is also the transmission of information. Biological information is stored in the one-dimensional form of a linear code (DNA, RNA) that, at the functional level, is translated into the three-dimensional structure of proteins. Prior to multiplication, the one-dimensional genetic code is copied, and complete sets of the code are transmitted to each of the two daughter cells that originate from binary fission. An analogous process occurs in minerals, where information may be stored in the two-dimensional lattice of a crystal plane. If a mineral has a strong preference for cleaving across the direction of growth and in the plane in which the information is held (Cairns-Smith 1982), the information can be reproduced. Note that in contrast to living cells, information can be stored in a crystal in two or three dimensions. However, copying three-dimensional information would be very challenging. Another important question is whether the stored information has actual meaning. We know that DNA has meaning because of the expressed segments of DNA (exons) that are read by the molecular machinery of the cell. However, there is no obvious way of assessing whether any ion patterns in minerals have a meaning. The prerequisite of an information-containing structure is a large number of meta-stable states (MacKay 1986). Meta-stable states are plentiful in clay minerals. The cations of Mg, Al, Fe(II), Fe(III), Li, Ti, V, Cr, Mn, Co, Ni, Cu, Zn can be substituted in octahedral sites (Bailey 1986); Si, Al, and Fe(III) can be substituted in tetrahedral sites of clay minerals (Newman and Brown 1987). Even more possibilities of exchange exist in the interlayer sites (individual cations, hydrated cations, hydroxide octahedral groups, etc). We should not dismiss out of hand the possibility that cation position in minerals may have meaning since exchangeable cations in the octahedral or tetrahedral sites can control the position of other cations. For example, amesite is an aluminum-rich trioctahedral serpentine mineral based on a 1:1 layer that can occur in spiral form (Anderson and Bailey 1981; Fig. 2.1). Substitution of Al for Si in one of every two tetrahedral sites is compensated by a similar amount of substitution of Al for Mg in octahedral positions (Bailey 1986). To the extent that these variations in mineral structure affect the functional properties of the mineral, the structural organization can be said to have functional meaning.

Another hallmark of life is said to be adaptation to the environment. Adaptation can be achieved by an individual organism in a transient and reversible way, or by collective individuals through time by the essentially irreversible mechanism of natural selection. An example of adaptation at the cellular level within individual organisms is the process of induction. Enzyme induction consists of a series of processes by which cells produce enzymes specific to a particular substrate only when that substrate is present.

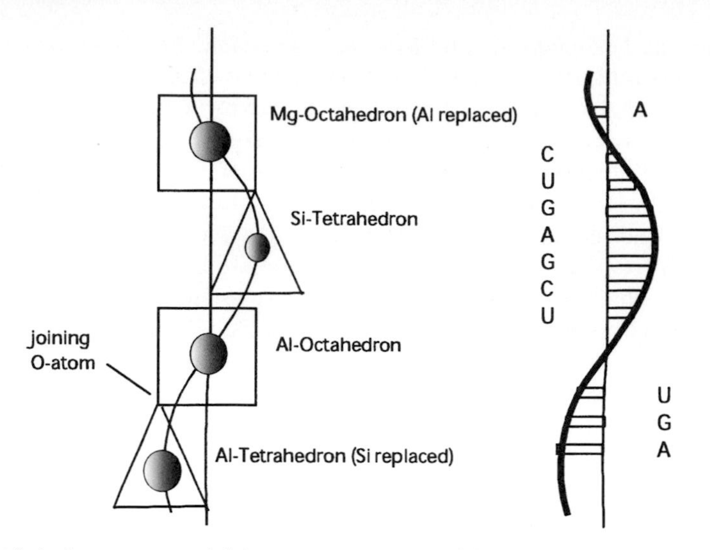

Fig. 2.1 Spiral structure of (a) amesite-$2H_2$ and (b) RNA, two-dimensional and schematic. As the sequence of nucleotides in RNA (C = cytosine, G = guanosine, U = uridine, A = adenosine) forms an information code, so could be in principal the cation order in this amesite.

Other examples of individual adaptive mechanisms include the adjustment of microbial size to nutrient conditions, or the movement of a cell to a nutrient source or away from toxic substances.

Genetic change from one generation to another can be initiated by a number of processes such as mutation or transposition of genetic material, transformation (e.g. DNA from a dead cell taken up by a living cell), conjugation (DNA transferred by means of quasi-sexual joining of cells), lateral transfer (DNA transferred from one living organism to another distantly related organism) and transduction (DNA transferred by viruses). Those changes that survive the filter of natural selection are regarded as genetic adaptations. However, clay minerals can also adapt to their surroundings. The most common form of adaptation for clay minerals with interlayer sites is their response to water availability. If the outside environment is dry, water is released from the interlayer to surroundings; if the outside environment is wet, water is adsorbed into the interlayer. Another adaptive characteristic of clays is the ability to accommodate different ions at ion exchange sites that will affect the lattice structure during clay mineral formation. Also, where chemical weathering is prevalent, as in the tropics, clays can develop an outer, weather-resistant layer of aluminum oxide or silicate. Changes in mineralogy can also occur depending on the ion source and change in the environment. Dehydration reactions involving water molecules or hydroxyl ions lead in general to a structural change. In extreme cases, one clay mineral can transform into another. For example, gibbsite can transform into boehmite because of an alternation of dry and wet periods (Boulange et al. 1997).

Adaptation can also be defined as "...changes originating internally" (Random House 1987). While minerals grow generally at their outside margins, which are exposed to the environment, chemical changes can also occur inside minerals, as when water molecules move into the interlayer of clays if water availability in the outside environment increases, or if the clays accommodate various ions at ion exchange sites. While this could be viewed as a simple internal response to an environmental change that originates externally, many homeostatic adjustments within living organisms could be similarly described. Thus, the distinction is at least semantically ambiguous.

A second dictionary definition of life is the cumulative aggregation of all the individual criteria listed in the first definition ("2. The sum of the distinguishing phenomena...") (Random House 1987). As indicated above, good analogies can be found for each of the traditional criteria for living systems in the inorganic world, specifically for clay minerals and metal oxides. A summary is provided in Table 2.1. While we are not aware of any specific minerals that display all four characteristics, there is no reason conceptually to assume that a mineral fitting the sum of all the criteria cannot exist. At the biological edge of the interface between the living and non-living world, viruses present a similar case in reverse. By the traditional definition viruses are not considered living entities because they cannot reproduce and grow by themselves and do not metabolize. Nevertheless, they possess a genetic code that enables them to reproduce and direct a limited amount of metabolism inside another living cell. They thus fulfill the traditional criteria only part of the time and under special circumstances. At best, by the traditional definition, they could be considered to be "reversibly alive". Since viruses presumably evolved from bacteria that clearly are alive, do they represent a case in which a living entity has been transformed to a non-living state by natural selection? If we accept the proposition that viruses are not alive, how would we consider parasitic organisms or bacterial spores? Parasites cannot grow by themselves either and spores remain in dormant stages with no dynamic biological attributes until they become active under favorable environmental conditions. Thus, if we consider parasites or bacterial spores to be alive, the logical consequence would be to consider viruses alive as well.

In summary, the traditional definition of life on close examination fails to distinguish consistently between the living and the non-living world (Table 2.1). Since biology and mineralogy have both been characterized extensively on Earth, the distinction between the two is not difficult to make on our home planet. Notwithstanding the semantic ambiguities, we generally know life (or non-life) when we see it. But the definition matters more when we leave the familiar context of Earth, and encounter more exotic conditions and environments where dynamic phenomena may exist with which we are unfamiliar. In that context, semantic ambiguities become conceptual stumbling

Table 2.1 Properties of life: organic mechanism and inorganic parallel (modified from Schulze-Makuch et al. 2002).

Property	Basic Requirement	Organic Mechanism	Inorganic Parallel
Metabolism	Yield of energy by electron transfer	Various types of biochemical pathways including photosynthesis; Energy storage as ATP or GTP	Energy uptake via heat or light, elevation of electrons to higher energy bands, where they can absorb energy at specific frequencies; Energy storage by luminescence or at interlayers of clay minerals due to high heat capacity of water and OH-groups
Growth	Increase in size of single unit	Cell growth as long as nutrients are available and environmental conditions are favorable until reproduction occurs; Self-organizing as development proceeds, errors corrected by enzymes	Crystal growth as long as favorable environmental conditions prevail and a sufficient ion source is present; Local surface reversibility makes it possible to correct certain mistakes during the growth of silicate or metal oxide crystals from aqueous solution
Reproduction	Multiplication of information	Various mechanisms of reproduction (most commonly by binary fission); The genetic code is preserved and duplicated between different generations	Crystals commonly break up during their growth because of cleavage; if a mineral has a strong preference for cleaving across the direction of growth and in the plane in which the information is held, new "individuals" may form and each of the new "individuals" may expose the information on the new surface (Cairns-Smith 1982)
Adaptation to Environment	Reaction to, compensation for, and development of new abilities in response to various types of environmental changes	Genetic adaptation from generation to generation: mutations, transposition of genetic material, transformation, conjugation, transduction; Homeostatic adaptation within individuals: movement to nutrient source, cellular shrinkage or formation of spores in nutrient-poor environment, enzyme induction	Adaptation from generation to generation: changes in mineralogy due to changes in environment and ion source (e.g. in extreme case transformation from one (clay) mineral to another) Adaptation within individuals: adaptation to outside environment via water release or adsorption in interlayer (clay minerals), development of outer, weathering-resistant layers such as Al_2O_3 or silicate layers in tropical soils

blocks and observational obstructions. For that reason, we need a definition of life that more effectively and precisely captures the fundamental essence of the phenomenon for which we are searching.

2.2 Modern Definitions of Life

As weaknesses in the traditional definition of life have become apparent, efforts to define life have become increasingly sophisticated and abstract, in an attempt to identify the essence of living systems that distinguish them from physical entities – both naturally occurring and fabricated – that are not alive (Margulis and Sagan 1995, Luisi 1998, Lahav 1999). Although a final consensus on a suitable definition of life has not been reached, the historical debate over the nature of life has resulted in an improved understanding of life's most fundamental features.

The modern attempt to redefine life in a more sophisticated way dates from Schrödinger's introduction of physical aspects such as energy states and entropy as the essence of what it means to be alive (1944). A similar theme was developed later by Szent-Györgyi (1972). Moreno et al. (1990) focused on the autonomous nature of life by describing it as an autonomous system capable of self-reproduction and evolution. Maturana and Varela (1981) also emphasized the process of self maintenance, or "autopoiesis", as the fundamental essence of life. Lwoff (1962) and Banathy (1998) emphasized the information processing properties of life, while Dyson (1999) in a similar vein defined life as a material system than can acquire, store, process, and use information to organize its activities.

Some authors have striven for a comprehensive definition that focuses more on the continuity of life through time. Monod (1971), for instance, combined the ecological, thermodynamic, and bioinformatic properties of life, but added the ambiguous concept of teleonomy (apparent purposefulness in living organisms). Another effort at comprehensiveness is the recent proposal of Koshland (2002) of seven pillars of life, which he designated as a program, improvisation, compartmentalization, energy, regeneration, adaptability, and seclusion. Regeneration counteracts thermokinetic degeneration, and seclusion introduces the important concept of a bounded environment.

Since all living forms operate within the constraints of environmental conditions and limitations, some authors have tried to incorporate an ecological perspective into their definitions. For instance, Feinberg and Shapiro (1980) proposed to redefine life as the fundamental activity of a biosphere – a highly ordered system of matter and energy characterized by complex cycles that maintain or gradually increase the order of the system through an exchange of energy with its environment. Chyba and McDonald (1995) also focused on the interactions of life with its environment when they postulated that life is a system that acquires nutrients from its environment, responds to stimuli, and reproduces.

These admirable attempts to include an ecological perspective illustrate one of the problems that has bedeviled historical attempts to define life: namely, the confusion between life as a historical process, and the features of matter that constitute the state of being alive at a given moment in time. While almost all definitions of life refer in some way to reproduction as an essential feature, at a given moment an organism may be alive but not reproducing (Margulis and Sagan 1995). Similarly, some authors insist that the capacity for Darwinian evolution is an essential feature of life, yet any single organism during its lifetime is clearly not undergoing evolution. Thus, the condition of "being alive" needs to be distinguished from the "properties of a living system". The distinction is more than a semantic technicality, if the search for life on other worlds depends on the definition of what is being searched for. While ultimately we must know that what we discover is a "living system" capable of self-perpetuation, at the moment when we first encounter it, we need more precise and practical criteria for judging whether or not it is "alive".

2.3 Our Definition of Life

The definition of life used in this monograph was originally advanced in Schulze-Makuch et al. (2002). Our definition comprises of three fundamental characteristics: life is (1) composed of bounded microenvironments in thermodynamic disequilibrium with their external environment, (2) capable of transforming energy and the environment to maintain a low-entropy state, and (3) capable of information encoding and transmission. This definition was developed to emphasize its practical use for the detection of life beyond Earth. In particular we were interested in specifying those properties of life that would suggest favorable habitats for exploration and that could be identified by remote sensing techniques now or at some time in the future. The reader is referred to Chap. 8 for practical consequences of our definition of life and to Chap. 9 on how to use it to detect extraterrestrial life. In this section we lay the theoretical groundwork for our definition and discuss its advantages and disadvantages.

2.3.1 Bounded Microenvironments in Thermodynamic Disequilibrium

A closed natural inorganic system, isolated from its surroundings, adheres to the second law of thermodynamics and moves spontaneously toward a state of maximum entropy. It also moves toward a minimum amount of free energy with the Gibbs free energy between reactants and products being zero at equilibrium conditions. Life, on the other hand, maintains a high free energy state. This enables it, first, to do work on its environment. Secondly, the entropy of living systems is low because they are highly organized compared to their environments (even though the second law of thermodynamics remains valid, as it applies to the macrocosm as a whole). Minerals fall ambiguously between these two extremes. They are highly organized and therefore have low entropy (Fig. 2.2). But in a natural system they generally move spontaneously toward a lower state of free energy. However, as previously discussed, some luminescent minerals can absorb energy that temporarily elevates them to a higher free energy state than their external environment. Lacking a permanent storage mechanism, however, the energy gain is generally soon dissipated.

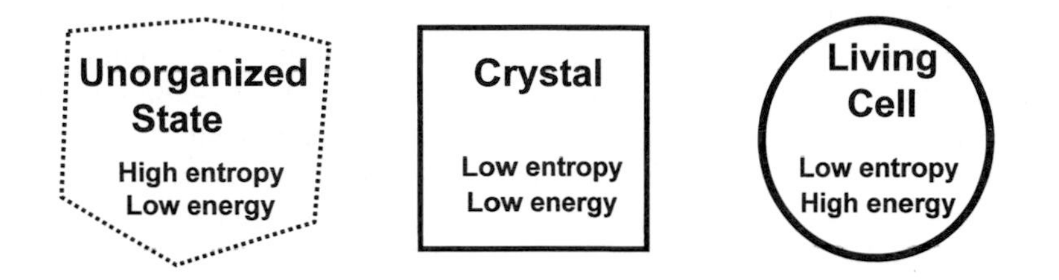

Fig. 2.2 Thermodynamic view of unorganized, crystalline and living state.

A major distinction between living and non-living systems is the presence of biomembranes. These establish boundaries that serve to (1) preserve the high free energy state of the system from dissipation, (2) encapsulate and confine a high concentration of interacting solutes and macromolecules, and (3) carry out complex functions such as selective solute permeation, light transduction, and the development of chemiosmotic potentials that generate energy gradients and provide the basis for reversible states of excitation (Deamer et al. 1994).

Disequilibrium on a cellular scale is made possible by the cell membrane, which enables the establishment of different solute concentrations within and outside the cell. On a supracellular scale disequilibrium conditions are created most visibly by colony-forming organisms such as stromatolites and corals, which are multicellular aggregates on a local scale (see microenvironments below). On a planetary scale, disequilibrium conditions can be established by biological processes such as photosynthesis.

As we criticized the traditional definition before by comparing it to analogs in the mineral world, it is only fair to evaluate how our definition stands up to these kinds of comparisons. Minerals do not have a membrane composed of fatty acids or similar compounds that living systems use. However, macromolecules within clay minerals can be protected. Clay particles can be linked in face-to-edge contact that results in an open internal framework with very high porosity (Bennett and Hulbert 1986). In such a clay fabric compartment, macromolecules (or clay minerals themselves) can be protected from disturbances in the environment. This type of framework structure can even lead to selectivity of specific ions. Some surface-active solids such as double-layer metal hydroxide minerals are capable of transporting matter against concentration gradients.

Stable disequilibrium conditions are also possible in the inorganic world but only to a limited extent (by sudden volcanic activity or for the time a hydrothermal vent is active). In a completely inorganic world plutonic rocks with their volatile components would be expected eventually to reach equilibrium with sediments, the atmosphere, and ocean water. Disequilibrium conditions or concentration gradients can be maintained within clay minerals for a period of time due to geometrical constraints and large energy barriers against ion exchange. Concentration gradients are most likely to form in the geometrically tight tetrahedral sites and to a lesser extent in the more spacious octahedral sites. However, lacking a distinctive boundary comparable to a biomembrane, clay minerals cannot maintain stable disequilibrium conditions at their reactive outer edges or in their reactive interlayer sites. Thus, like any inorganic system, this system would eventually reach equilibrium with its natural surroundings. Thus, while at any given instant disequilibrium can be achieved by inorganic processes, this condition cannot be maintained indefinitely. By contrast, living systems are able to establish order within a chaotic world and perpetuate that order as long as energy is available for resisting the inexorable tendency toward increased entropy.

2.3.2 Transformation of Energy and Environment to Maintain a Low Entropy State

A primary property of a living system is its ability to transform energy from external sources for the purpose of maintaining a low entropy state and performing work. While energy transformations are characteristic of all dynamic physical and chemical systems, energy flow in nature tends to result in greater disorder among all elements of the system. Energy released through different stages of the rock and water cycles, for instance, generally erodes land and distributes water to increase the entropy of the total collection of water and land toward equilibrium (lower mountains, more dispersed water and soil). The energy transformations of living systems, on the other hand, serve primarily to harvest and store the levels of free energy necessary for maintaining the highly ordered structure of the organism and performing the work that living cells carry out. The net effect for living systems, in contrast to that for non-living systems, is to maintain and often increase order at local levels and on microscopic scales.

There are two consequences to the way in which life transforms energy. One is that much of the energy is used to create and sustain a level of complexity that supports emergent functions that in their totality exceed the sum of the parts of the system. A mountain may be structurally complex but its role in the rock cycle is not dependent on the detailed organization of its individual rocks and sediments. The mountain is in essence a simple conglomerate of its component parts. The function of a living organism, on the other hand, depends critically on precisely how it is put together. Its component parts function in a coordinated manner, to generate a complex array of emergent properties, both structurally and functionally. The generation and maintenance of this complexity is one of the primary uses of the energy that living systems transform.

A second consequence of biological energy transformations is to create one or more additional microenvironments within the natural environment. The Eh (redox-potential), pH, solute composition, and structural complexity of the living cell is maintained at levels different from the extracellular environment because of the autonomous functions carried out by the cell, but not in the abiotic environment surrounding the cell. New environments can also be created on a larger scale by colony-forming organisms such as stromatolites and corals, which can alter the topography of large amounts of habitat. Life-induced changes can occur even on a planetary scale, such as the change in atmospheric oxygen composition brought about by oxygen producing microbes on Earth, beginning with the emergence of photosynthesis as a uniquely biological form of energy transformation (Schopf 1994, Knoll 1999). This innovation enabled life to become autotrophic (manufacturer of its own food from the simple and abundant molecule, CO_2) on a global scale. Thus, not only is the transformation of energy a characteristic of life, but so is the ability of life to alter conditions in the natural environment.

Stromatolites provide a particularly relevant example of how living organisms can alter their environment. On the early Earth CO_2 was more abundant in the atmosphere than it is today (Kasting and Brown 1998), so the CO_2 partial pressure was higher. Thus, CO_2 was more available in ocean water than today, which resulted in H_2CO_3 and HCO_3^- as the dominant carbon species in the water, with Ca^{2+} being present in soluble form. Stromatolites (algal mats produced by a type of cyanobacterium, Fig. 2.3) consumed the CO_2 from the water in shallow seas through photosynthesis. This resulted in an increase of carbonate in the sea water and the precipitation of calcium carbonate. Geological deposits called stromatolitic microbialites were formed as $CaCO_3$ was precipitated and sediments were trapped within the sticky layer of mucilage that surrounds the bacterial colonies. As this process occurred repeatedly, new layers of sediments were created to form the characteristic stromatolitic microbialites. Thus, the stromatolites modified their surrounding environment not only by changing the chemistry of the ambient seawater but also by altering the topography and mineral composition of the beach. A similar point can be made about corals that create large structures such as the Great Barrier Reef of Australia, which stretches more than 1000 km in length. Other well-known examples of how microbes affect their environment are the banded-iron formation of the early Earth, and the sulfur deposits in the Cyremaican lakes in Libya (Ehrlich 1981).

Microenvironments can also be created by inorganic processes. The zoning of crystals either during growth or decay processes such as weathering is a typ-

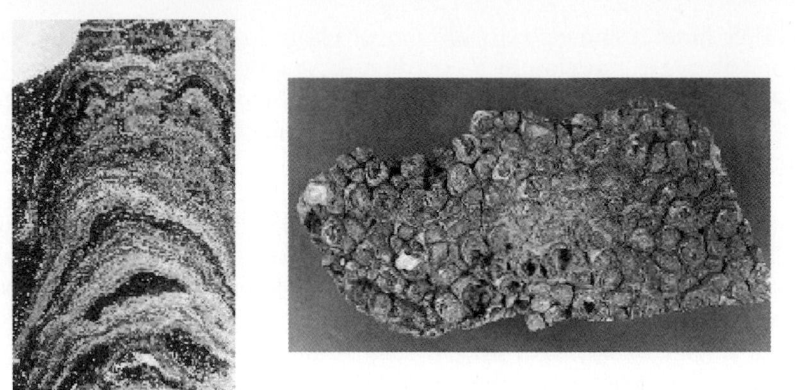

Fig. 2.3 Stromatolites, (a) cross-sectional view (340 m.a.; Five, Scotland; provided by Neil Clark, Curator of the Palaeontology Hunterian Museum of the University of Glasgow); (b) bird's-eye view of stromatolite (800 m.a., Amadeus Basin of Central Australia; provided by Stefan Arndt of the Institut für Spezielle Botanik, FSU Jena, Germany).

ical example. However, microenvironments created by inorganic processes are consistent with thermodynamic equilibrium processes. Minor exceptions to this rule are merely due to geometrical constraints such as the formation of boundary layers of a pH different from the ambient that can form in zeolite minerals and double-layer metal hydroxides via structural hydroxyl ions. Thus, an intrinsic property of life is that it will alter its environment in a way that is inconsistent with thermodynamic equilibrium processes.

2.3.3 Information Encoding and Transmission

A property unique to living systems is the transmission of information from one living entity to the succeeding entity. At the cellular level, this means that multiple descendent cells acquire the genetic information previously held by a single parental cell. At the level of the multicellular organism, it means that all the information for the organism's development and function is replicated, then passed through reproductive cells to the offspring (next generation). When a cell or organism can no longer maintain steady disequilibrium conditions it approaches equilibrium with its environment and therefore dies. Despite the death of the parent organism, the informational blueprint for the organism's structure and function, which for life as we know it on Earth is based on a chemical code, survives to the descendent organism, and will be transmitted from generation to generation. Despite the demise of the individual organism that harbors the code in its cells during any single generation, the transmission of the code to a subsequent generation ensures that the instructions for life specific to that particular kind of organism will persist.

Genetic codes are meaningful because of the characteristics that they impart to the systems in which they reside. However, a code of information by itself is useless if there is no meaning or consequence associated with it. The distribution of atoms in mineral lattice may be understood as a code, in that the information content of the distribution pattern is greater than zero, but there is no apparent meaning associated with it since there is no functional consequence to the pattern. Another main factor that distinguishes living systems from the mineral world is a sharp difference between substance and information. Genetic information is chemically codified in separate units (nucleic acids) of the cell that are distinct from their physical manifestation, hence the functional consequences of the code. A segment of DNA codes for a protein that carries out a particular function. If the protein is broken down, the function ceases, but the DNA that coded for the protein persists through replication and the reproduction of successive generations indefinitely. A mineral, on the other hand, may be capable of rearranging atoms and molecules in response to environmental influences such as weathering; but even if it is supposed that this rearrangement has changed the nature of information encoded in the mineral, the altered information has no effect apart from the specific rock in which it is embedded. Therefore, the information persists only so long as the mineral itself remains intact. There is no obvious consequence to the information, and no expression of it distinct from itself.

2.4 Implications for the Origin of Life

Based on our definition, a system has to do three things to be considered alive. It has to (1) construct and maintain a boundary to separate it from its environment, (2) transform some type of free energy into energy-yielding chemical reactions to sustain itself, and (3) replicate and transmit genetic information. We discuss those parameters and how they were probably exhibited for the first living forms on Earth. We initially focus on the origin of life on Earth because it is the only example we have, but we will attempt to extrapolate to other possible sites for the origin of life in the universe.

2.4.1 Inferences for the First Cellular Membranes

The cells of all living organisms on Earth use a selectively permeable membrane to preserve the high free energy state of the system from dissipation, encapsulate and confine a high concentration of interacting solutes and macromolecules, and carry out complex functions such as selective solute permeation, substrate interaction, and energy transduction. The core of these membranes is made from amphiphilic lipids such as fatty acids and phospholipids, in which the hydrophilic ends of the molecules are oriented toward the outer aqueous environment, while the hydrophobic ends are pointed toward the inside of the membrane core.

The formation of vesicles from amphiphilic molecules appears to be relatively easy. Deamer and Pashley (1989) extracted organic material from meteorites that formed cell-like membranes. The spontaneous formation of vesicles appears to be characteristic of phospholipid compounds and similar materials. There are a variety of suggestions as to how the first vesicles could have formed to become the precursor of the cell. Chang (1993) proposed bubble formation and breaking in the ocean–atmosphere interface as likely mechanisms for closing vesicles. Russell and Hall (1997, 2002) suggested the formation of iron sulfide membranes, precipitated as bubbles at submarine hydrothermal vents, as primordial cell membranes in a step leading to life. Also recently, Deamer et al. (2002) pointed out that amphiphilic molecules having carbon chains with lengths greater than six carbons form micelles as concentrations increase above a critical value. At chain lengths of eight carbons and higher, bilayers begin to appear in the form of membranous vesicles, which become the dominant structure as concentrations increase further (Fig. 2.4).

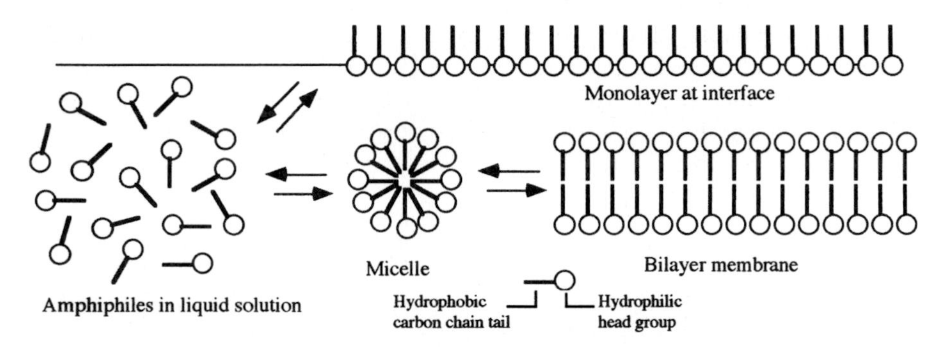

Fig. 2.4 Self-assembled structures of amphiphiles (from Deamer et al. 2002).

Phospholipids appear to be ideally suited as cell membrane constituents for life on Earth. That does not limit the possibilities though. In principal, many amphiphilic molecules could serve as a cellular boundary. Conditions different in temperature, pH, and redox-potential of the liquid medium could favor amphiphilic compounds other than phospholipids. Amphiphilic compounds with outwardly oriented polar groups are well adapted for separating internal and external compartments submerged in a polar solvent. If the liquid environment were hydrophobic rather than hydrophilic, as likely on Titan, polar head groups at the surface of the membrane would be repelled by the hydrophobic solvent. How would a membrane be constructed in such an environment? Would molecules consisting of polar groups flanked by hydrophobic chains provide effective compartmental barriers in an organic milieu? Empirical studies are needed to answer this question. Since Titan may well represent a prototype of certain exotic environments found on other worlds, the search for membranes of a different composition from those known for terrestrial organisms should be a priority for laboratory research in astrobiology.

2.4.2 Inferences with Regard to the First Metabolism

The chemical reaction that supported the first life on Earth must have been simple but very efficient. A reaction that appears to fit this description is the oxygenation of hydrogen to water. The oxidation can occur with oxygen or some other oxygenated compound:

$$H_2 + \frac{1}{2}O_2 \longrightarrow H_2O \tag{2.1}$$

Reaction (2.1) yields an energy of 237.14 kJ/mol (or 2.6 eV) per reaction. There are various lines of evidence for why this reaction has such a central meaning: (1) hydrogen is the most common element in the universe, and gaseous hydrogen molecules are supplied by volcanic outgassing and by the abiotic reaction of water with basaltic rocks (Stevens and McKinley 1995), so are readily available, (2) oxygen or oxygenated compounds are common in the universe and also readily available for reaction, and (3) close relatives of the most ancient microbes on Earth use this reaction for metabolism. On Earth, where volcanic activity, and with it carbon dioxide, are common, this reaction is often coupled with the reduction of CO_2 and called methanogenesis:

$$4H_2 + CO_2 \longrightarrow CH_4 + 2H_2O \tag{2.2}$$

The reduction of CO_2 is energetically unfavorable but proceeds because of the net energy gain by the oxidation of 2 hydrogen molecules to water. The net energy gain of this reaction is 130.63 kJ/mol or 1.4 eV (474.28 kJ/mol of energy gained by the oxidation of two hydrogen molecules to water minus the 343.65 kJ/mol lost by the reduction of carbon dioxide to methane).

This kind of reaction can set the stage for a simple feedback process or pre-ecosystem. If the methane produced is exposed to oxygen-rich conditions, for example in some oceanic layer, it would be favorable to oxidize the methane back to carbon dioxide with an energy gain of 6.1 eV to close the loop for a nutrient cycle. However, the H_2 would not return to its molecular state, but most likely would be incorporated into some other chemical compound. Thus, the limiting compound in this feedback cycle is the supply of molecular hydrogen.

As an alternative, hydrogen may be oxidized in a simple but efficient reaction with the help of sulfur or iron as shown in (2.3) and (2.4), respectively.

$$H_2 + S \longrightarrow H_2S \tag{2.3}$$

$$H_2 + 2Fe(III) \longrightarrow H^+ + 2Fe(II) \tag{2.4}$$

The energy yield from reaction (2.3) is 33.4 kJ/mol (0.36 eV per reaction) and 148.6 kJ/mol (1.6 eV per reaction) for (2.4). Both of these reactions provide sufficient energy to form energy-storing compounds (the terminal phosphate bond in ATP, a later development of high-energy storage efficiency, needs only about 0.3 eV per molecule to form). Iron and especially sulfur are readily available in volcanically active regions, thus the arguments made above for (2.1) can be made equally strong for reactions (2.3) and (2.4). The major advantage of oxidation of hydrogen via oxygen is its superior energy yield. However, on early Earth molecular oxygen was very rare. Also, in an early development state this reaction may have been too energy-rich, and the more benign oxidation of hydrogen to hydrogen sulfide may have been more controllable and suitable. A similar feedback mechanism as elaborated above for the oxidation of hydrogen via oxygen can be established for sulfur, iron, and other compounds (Schulze-Makuch 2002). This indicates that the use of chemical energy, by itself or as a feedback cycle allowing microbial differentiation, should be fairly easy to establish. Photosynthesis, on the other hand, is a much more complex mechanism with many intermediate steps, and because of its complexity probably evolved after chemoautotrophy. Also, photosynthesis requires the conversion from light energy back to chemical energy (to build energy storage compounds such as ATP, cellular components, etc); which adds another layer of complexity. Hose et al. (2000) found that green sulfur bacteria that used photosynthesis outside a cave oxidized hydrogen sulfide to elemental sulfur inside the lightless cave environment. This finding could support a hypothesis that photosystem I (simplified in (2.5)) may have developed from the chemotrophic oxidation of hydrogen sulfide, and that the observed green sulfur bacteria are capable of switching back to the older chemotrophic metabolic pathway when needed. Other opinions exist, however. Hartman (1998), for example, proposed photosynthesis as the key to the origin of life.

$$2H_2S + CO_2 + \text{light} \longrightarrow CH_2O + H_2O + S_2 \tag{2.5}$$

Based on the previous discussion, it appears that the oxidation of hydrogen coupled with the reduction of some other compound is the most fundamental metabolic pathway for chemotrophic life. However, it does not mean that the first organisms were necessarily based on this type of metabolism. The oxidation of hydrogen does not occur under temperature and pressure conditions on Earth's surface without very good catalysis. Chemoautotrophic organisms have more complex internal requirements than, for example, heterotrophs feeding on organic macromolecules (McClendon 1999). Thus, heterotrophic organisms that gobbled up high-energy organic molecules present on the primordial Earth are a feasible scenario for the first cells on Earth. However, given the abundance of hydrogen and carbon in the universe, and the pervasiveness of volcanic activity, at least in our solar system, it seems reasonable to infer that life elsewhere would also be based on chemoautotrophy; if not initially, then shortly after organic molecules become scarce (not being replenished at a high enough rate), and that chemoautotrophic life would probably oxidize hydrogen, possibly combined with the reduction of carbon dioxide.

2.4.3 Inferences for the First Replication Mechanism

Replication is the ultimate challenge for origin of life researchers. The pre-evolutionary emergence of mechanisms for replication of genetic information are difficult to imagine, primarily because of the involvement of proteins whose structure presupposes a coding mechanism. Benner (2002) pointed out that a suitable biopolymer would have to be able to replicate, catalyze, and evolve without a loss of properties essential for replication, which is achieved in DNA and RNA by repeating charges. Since RNA has been shown to be capable of some catalytic activity (Guerrier-Takada and Altman 1984, Cech 1985, Ferris 1993, Schwartz 1993, Lazcano 1994), in principle it could have served as the first template and catalyst for its own replication. However, the RNA found in contemporary cells is quite labile to environmental degradation. If that was the case at the advent of life, before ribonuclease enzymes were common, RNA would not have been plausible as either the first replication template or agent. Furthermore, in order to be a suitable replicator under early terrestrial conditions, RNA molecules would have to have been very simple and capable of reproducing at a very low error rate. It is not clear whether RNA could have fulfilled this requirement. What was needed was a biopolymer that could be reproduced autocatalytically and undergo chemical evolution. Lipids, peptide nucleic acids (PNA), threose nucleic acids (TNA), peptides, and proteins have also been considered as the basis for the origin of life on Earth (Nielsen 1993, Orgel 1998, Schöning et al. 2000, Bada and Lazcano 2002, Chaput and Szostak 2003). In the view of many theorists, a prior stage in the evolution of life – an intermediate link long since lost in the evolutionary progression toward the current sophisticated mechanism – is needed to explain how replication was first achieved.

The difficulty of identifying the first replicator led Bernal (1967) to propose the idea that clays or minerals were actively involved in the origin of life. This appears reasonable on the grounds that many properties of life can also be exhibited by minerals, especially clay minerals and metal hydroxides, as discussed before (e.g. Table 2.1). Cairns-Smith (1982, 1985) suggested catalytic clay surfaces as the first genetic mechanisms. A variety of biogenic compounds have been observed to interact with clays such as fatty acids, sugars, amino acids, and proteins. Ferris (1993) even observed the formation of RNA oligomers on montmorillonite. Huber and Wächtershäuser (1998) have shown the synthesis of peptides on a (NiFe)S surface. Wächtershäuser (1994) and Huber and Wächtershäuser (1998) have proposed a sequence of mineral catalysts, which spontaneously form positive feedback synthesis involving both organic and inorganic compounds. It is interesting to note that adsorption and binding of DNA on clay and sand particles protect the DNA against degradation by nucleases without inhibiting its transforming ability (Lorenz and Wackernagel 1987, Khanna and Stotzky 1992, Paget et al. 1992). RNA can establish an even stronger interaction with clay than DNA. Single-stranded RNA, in fact, can interact with the clay substrate, not only by the formation of hydrogen bonds between the phosphate groups of the RNA and the silanol or octahedral Al(III) groups of the clay, but also through its nitrogen bases. The same would be much more difficult for a double-stranded molecule such as DNA with its nitrogen bases inside the double helix (Franchi et al. 1999). This adds further circumstantial evidence that RNA came before DNA and that mineral surfaces are in some way involved in the first replication mechanism. DNA and RNA both use complementary rather than identical components for replication. A replicator copying identical components seems to be more simplistic and the precursor of RNA and DNA may have functioned that way. It should be pointed out that chemical coding of information is not the only possibility. In principle, any type of stored information that can be replicated and transmitted from one generation to the next would be workable (for a proposed alternative mechanism on replication with magnetic orientations, see Fig. 8.1).

2.4.4 Other Inferences

2.4.4.1 Size

Although life can certainly be macroscopic (we are evidence of it), life's origin must have occurred microscopically. Bounded microenvironments must have a surface-to-volume ratio that allows diffusion throughout the cell in a brief period of time. Anything larger would slow the rate of essential metabolic reactions to a presumably non-viable level and require large amounts of energy to maintain the structural integrity of the cell. At the same time, a cell has to be large enough to host the molecular machinery required for carrying out its metabolic and reproductive functions. While eukaryotes developed internal specializations such as membrane-bounded organelles to enable growth to larger sizes, the first organisms were surely simpler, undifferentiated, and correspondingly minute. While the revolutionary fusion of two or more prokaryotic cells apparently led to the subcellular specialization that enabled eukaryotes to assume an increase in cellular size by an order of magnitude (Margulis and Sagan 1995), the further revolutionary innovation of multicellularity was required to bring about the emergence of macroscopic organisms (Cowen 1995).

It has also been argued that there is a lower as well as upper limit to size. Schrödinger (1944) discussed why we observe organisms the size they are. The disordered activity of thermal motion is too prone to generate random deviation from the statistical determinism upon which a complex set of interactions depends, if the number of interacting components is not very large. A cell too small to contain a sufficient number of reactants would thus be subject to the risk of failure due to "sampling errors" in the normal course of its chemical and physical activities. However, convincing empirical confirmation of a definite lower limit for the size of a viable cell is not available, and evidence is accumulating on nano-sized about 100-fold smaller than common bacteria. For example, 100–200 nanometers (nm) large bacteria were identified in mammalian cells (Ciftcioglu and Kajander 1998, Kajander et al. 2001) and 400 nm hyperthermophilic archaea from a submarine hot vent (Huber et al. 2002). However, Schieber and Arnott (2003) interpreted nanobacteria as by-products of enzyme-driven tissue decay. This discussion is particularly significant in the light of claims that particles in the Martian meteorite ALH48001, which are much smaller than conventional terrestrial bacteria, could represent the remnants of life on Mars (McKay et al. 1996).

2.4.4.2 Type of Medium

The presence of a liquid medium is usually assumed for the origin of life on Earth, and indeed the presence of a liquid medium is very favorable. A liquid can provide a suitable medium in which atoms and molecules can move around relatively freely, encounter their reaction partners, and undergo chemical reactions in a reasonable time frame. In a solid medium atoms are essentially fixed in place, and each atom can only react with its immediate neighbors, limiting severely the possibility of complex reactions. In the gas phase densities of atoms and molecules are usually low, and time periods between collisions and interactions between various constituents are large. Thus, complex molecules are not created or transformed in a reasonable time frame (before their disintegration). A liquid has the additional advantage that molecules present in some liquids such as water dissolve into charged ionic species, enhancing reaction rates by orders of magnitude. A liquid medium also allows easy transport of nutrients and disposal of wastes, although some nutrients such as N_2 or CO_2 can also be assimilated from the atmosphere by Earth organisms. Thus, under conditions currently prevailing on Earth, a liquid medium is clearly essential for living processes (for a more comprehensive discussion, see Chap. 6). However, on other planetary bodies a gas can be compressed by gravity or other forces to a similar or higher density than a liquid, thus making up some of the disadvantage compared to a liquid medium. A different scenario can be envisioned in a gaseous atmosphere where no ozone shield is present to drastically reduce the penetration of UV radiation and/or where the atmosphere is relatively thin to allow more ionizing radiation to penetrate. In that case, ions, radicals, and electrons are created that are highly reactive, producing versatile chemical species (to some extent this is the case in Titan's atmosphere). Feinberg and Shapiro (1980) suggested that dense gases at high reaction-enhancing temperatures could be an equally convenient medium for chemically complex reactions and one which is much more common in the universe than liquid media. Sagan and Salpeter (1976) went so far as to envision specifically adapted organisms living in Jupiter's dense atmosphere. Evaluating the merit of these ideas is complicated by the fact that no distinct boundary exists between the liquid and gaseous medium at high temperatures and pressures. For example, water reaches the supercritical state at temperatures above 400 °C and pressures above 200 bar, at which conditions it cannot be described adequately as either liquid or gas. While empirical observations suggest that life tends to thrive at boundaries between states of matter, there appears to be no theoretical obstacle to the origin and persistence of life in the absence of such boundaries.

2.4.4.3 Environmental Conditions

Predictions of how life could have originated and on what planetary bodies life could be expected to exist would be greatly enhanced if the environmental conditions under which life can form would be known. However, even on Earth we are uncertain of the conditions under which life originated. Many researchers (e.g. Kompanichenko 1996, Stetter 1998) favor a hyperthermophilic origin of life based on the abundance of organisms discovered at hydrothermal vents on the ocean floors and the results from molecular biology, which appear to indicate that the universal tree of life is rooted in hyperthermophiles. However, high temperatures are generally detrimental to organic synthesis reactions, and the hyperthermophilic last common ancestor could have been simply a deep sea survivor from a near – sterilizing meteor impact. Others have proposed an origin for life on Earth by panspermia, the transport of living forms to Earth from an extraterrestrial source (Arrhenius 1903, Crick and Orgel 1973, Hoyle 1983). The exchange of viable microorganisms between planets in our solar system appears to be possible via exchange of meteoritic material (Davies 1996); however the exchange between stellar systems is statistically so unlikely that the origin of life on Earth must be sought within the confines of our solar system (Melosh 2003). Trace element composition in bacteria, fungi, and higher organisms shows a strong correlation with the concentrations of these elements in sea water (Goldsmith and Owen 2001), which would support a terrestrial origin of life. Recently, (Deamer et al. 2002) proposed a plausible scenario for the origin of life in an aqueous environment with a moderate temperature ($< 60\,°C$), low ionic strength, and pH values near neutrality (pH 5–8) with divalent cations at submillimolar concentrations. These conclusions were derived because the high salt concentration of the present oceans would exert a significant osmotic pressure on any closed membrane system, and divalent cations such as Ca^{2+} and Fe^{2+} (Fe^{2+} present in the ocean because of the absence of atmospheric oxygen on early Earth) would have a strong tendency to bind to anionic head groups of amphiphilic molecules inhibiting them from forming stable membranes (Monnard et al. 2002). Only a better understanding of the environment in general, the possible habitats for the earliest life, and the processes operating on the early Earth can lead to better insights into in which type of environment the first assembly of a living cell took place.

2.5 Implications for the Remote Detection of Life

If life conforms to the way we have defined it, our efforts will be maximized by focusing on the consequences of and requirements for the three components of our definition.

If life consists of bounded microenvironments in disequilibrium with their environment, we will be seeking to detect evidence of very small entities (in relation to the resolving power of our instruments), though aggregates of those entities, analogous to stromatolite mats and coral reefs on Earth, may be large enough collectively for detection. The necessity for enclosure in a barrier, most likely and often in a liquid medium, leads us to look for particular types of molecules in particular environments – probably amphiphilic molecules. Solvents other than water may be used by life elsewhere, which is discussed in Chap. 6. Also, any evidence of local chemical concentrations or physical properties distinct from their surroundings would be presumptive evidence for the possibility of life.

The nature of energy transformations – a nearly universal component of all definitions of life – depends on the nature of the energy gradients available. Since the easiest and most obvious way to detect biologically driven energy transformations is probably through either localized or global effects of the transformation, our instruments need to be attuned to the forms of energy available on the planetary object of our search. Energy sources other than light or chemical energy may be used by life elsewhere, which is discussed in Chap. 4. And because of the simplicity of chemical conversions as a source of energy, disequilibrium chemistry in the atmosphere or global habitat is probably relevant anywhere.

The third component of our definition – capacity for information encoding and transmission independent of the life span of the individual organism – will be virtually impossible to confirm until actual samples of the candidate life forms are in hand. The constraints applied by this part of the definition for remote detection strategies are thus limited. However, this part of the definition becomes critical once candidate samples are available, for only this third component will enable confirmation that the candidate structure in question is or was alive. The validity of this point is illustrated by the fact that all the lines of evidence in support of fossil organisms in the Martian meteorite ALH84001, including the pictures of nanomicrobial-like organisms (McKay et al. 1996), are inconclusive, absent confirmation that the fossil actually derived from an organism that did in fact replicate essentially identical copies of itself.

In Chap. 9, we discuss the technical approaches needed for detecting life as we have defined it on other worlds.

2.6 Chapter Summary

The definition of life is a problem that the brightest scientists have wrestled with for centuries but have not yet resolved. There are several obstacles to achieving a universally acceptable definition: (1) living systems use compounds that are abundant in the surrounding environment and processes that are not intrinsically different from reactions that occur inorganically, (2) there does not appear to exist a single characteristic property that is both intrinsic and unique to life, (3) there was probably no sharp line but rather a transitional boundary between a non-living and a living state of matter, (4) the condition of "being alive" has to be distinguished from the "properties of a living system" (i.e. an organism can be alive but does not reproduce or undergo evolution at any given moment). The definition of life that we use here was designed as a practical guide for anticipating the conditions on other worlds that might be suitable for the indigenous origin of life, and to assist in detecting life elsewhere: we propose that life is (1) composed of bounded microenvironments in thermodynamic disequilibrium with its external environment, (2) capable of transforming energy and the environment to maintain a low entropy state, and (3) capable of information encoding and transmission. From these basic assumptions of what life is, various inferences can be drawn about its origin and the type of habitats in which it might be found. First, the most primitive organisms must have been microbial in size. Metabolic reactions have to occur at a reasonable rate for a cell to stay viable. The most efficient size is a cell with microbial dimensions, not too large to waste energy on structural integrity or provide insufficient surface-to-volume ratios, but large enough to host the molecular machinery required for carrying out its metabolic, reproductive, and functional needs. Remote detection from orbiters would therefore probably be restricted to visualizations of large aggregates of such organisms, while landers would need to be equipped with microscopes for a realistic view of individual organisms. The first forms of life probably derived their energy either by oxidizing hydrogen or consuming organic, energy-rich molecules. Any other world on which the chemical capacity for oxidizing hydrogen existed would be a candidate for the origin of life and its persistence to the present day. In fact, any world on which an energy gradient exists is potentially capable of supporting life. Finally, any genetic program for the encoding and transmission of information must have been based on some form of macromolecular heterogeneity, possibly in concert with inanimate scaffolds such as clay, prior to the advent of RNA. To the extent possible, then, the search for life on other worlds must ultimately identify a coding mechanism, chemical or otherwise, to confirm that the candidate form of life is or has in fact been alive. While these inferences may be somewhat biased by the one case of life that we know – on Earth – we would close this chapter by noting that the composition of living matter resembles the composition of stars more closely than the composition of our planet, so terrestrial life may be more typical for life in the universe than we think.

References

Anderson CS, Bailey SW (1981) A new cation ordering pattern in amesite-$2H_2$. *American Mineralogist* 66: 185–195.

Arrhenius S (1903) Die Verbreitung des Lebens im Weltenraum. *Umschau* 7: 481–485.

Bada JL, Lazcano A (2002) Some like it hot, but not the first biomolecules. *Science* 296: 1982–198.

Bailey SW (1986) Layer silicate structures. In: Cairns-Smith AG, Hartman H (eds) *Clay minerals and the origin of life*, Cambridge University Press, Cambridge, UK, pp 24–40.

Banathy BA (1998) An information typology for understanding living systems. *Biosystems* 46: 89–93.

Benner S (2002) Weird life: chances vs. necessity (alternative biochemistries). Presentation given at "Weird Life" Planning Session for National Research Council's Committee on the Origins and Evolution of Life, National Academies of Sciences, USA, http://www7.nationalacademies.org/ssb/weirdlife.html.

Bennett RH, Hulbert MH (1986) *Clay microstructure.* International Human Resource Development Corporation, Boston.

Bernal JD (1967) *The origin of life.* World Publ., Cleveland.

Boulange B, Ambrosi J-P, Nahon D (1997) Laterites and bauxites. In: Paquet H, Clauer N (eds) *Soil and Sediments: Mineralogy and Geochemistry*, Springer-Verlag, Berlin.

Cairns-Smith AG (1982) On the nature of primary genetic materials. In: Cairns-Smith AG (ed) *Genetic takeover*, Cambridge University Press, Cambridge, UK, pp 136–163.

Cairns-Smith AG (1985) *Seven clues to the origin of life.* Cambridge University Press, Cambridge, UK.

Campbell NA (1996) *Biology.* Benjamin/Cummings, Menlo Park, California.

Cech TR (1985) Self-splicing RNA: implications for evolution. *Int. Rev. Cytol.* 93: 3–22.

Chang S (1993) Prebiotic synthesis in planetary environments. In: Greenberg JM, Mendoza-Gomez CX, Pirronello V (eds) *The Chemistry of Life's Origins.* Kluwer Academic Publishers, Dordrecht, pp 259–300.

Chaput JC, Szostak JW (2003) TNA synthesis by DNA polymerases. *J. Am. Chem. Soc.* 125: 9274–9275.

Chyba CF, McDonald GD (1995) The origin of life in the solar system: current issues. *Ann. Rev. Earth Planet. Sci.* 23: 215–249.

Ciftcioglu N, Kajander EO (1998) Interaction of nanobacteria with cultured mammalian cells. *Pathophysiology* 4: 259–270.

Cowen R (1995) *History of life.* Blackwell, Boston.

Crick FHC, Orgel LE (1973) Directed panspermia. *Icarus* 19: 341–345.

Curtis H, Barnes NS (1989) *Biology.* Worth, New York.

Davies PCW (1996) The transfer of viable microorganisms between planets. Ciba Foundation Symposium 202 (Evolution of hydrothermal ecosystems on Earth (and Mars?)) Wiley, Chicester.

Deamer DW, Pashley R (1989) Amphiphilic components of the Murchison carbonaceous chondrite: surface properties and membrane formation. *Orig. Life Evol. Biosphere* 19: 21–38.

Deamer DW, Harang-Mahon E, Bosco G (1994) Self-assembly and function of primitive membrane structures. In: Bengtson S (ed) *Early life on Earth*, Columbia University Press, New York, pp 107–123.

Deamer D, Dworkin JP, Sandford SA, Bernstein MP, Allamandola LJ (2002) The first cell membranes. *Astrobiology* 2: 371–382.

Dyson F (1999) Life in the universe: is life digital or analog? Abstract of a Scientific Colloquium given 3 December 1999 at the Goddard Space Flight Center.

Ehrlich HL (1981) Geomicrobial transformations of sulfur. In: Ehrlich HL (ed) *Geomicrobiology.* Marcel Dekker Publishing Company, New York, pp 251–280.

Feinberg G, Shapiro R (1980) *Life beyond Earth – the intelligent Earthling's guide to life in the universe.* William Morrow and Company, Inc, New York.

Ferris JP (1993) Prebiotic synthesis on minerals: RNA oligomer formation. In: Greenberg JM, Mendoza-Gomez CX, Pirronello V (eds) *The chemistry of life's origins.* Kluwer Academic Publishers, Dordrecht, pp 301–322.

Franchi M, Bramanti E, Bonzi LM, Orioli PL, Vettori C, Gallori E (1999) Clay-nucleic acid complexes: characteristics and implications for the preservation of genetic material in primeval habitats. *Orig. Life Evol. Biosphere* 29: 297–315.

Goldsmith D, Owen T (2001) *The search for life in the universe.* Benjamin/Cummings Publishing Company, Menlo Park, California.

Guerrier-Takada C, Altman S (1984) Catalytic activity of an RNA molecule prepared by transcription in vitro. *Science* 223: 285–286.

Hartman H (1998) Photosynthesis and the origin of life. Orig. Life Evol. *Biosphere* 28: 515–521.

Hazen RM (2002) Emergence and the origin of life. In: Palyi G, Zucchi C, Caglioti L (eds) *Fundamentals of Life.* Elsevier, Paris, pp 277–286.

Hose LD, Palmer AN, Palmer MV, Northup DE, Boston PJ, DuChene HR (2000) Microbiology and geochemistry in a hydrogen-sulphide-rich karst environment. *Chemical Geology* 169: 399–423.

Hoyle F (1983) *The intelligent universe.* Michael Joseph, London.

Huber C, Wächtershäuser G (1998) Peptides by activation of amino acids with CO on (NiFe)S surfaces. *Science 281*: 670–672.

Huber H, Hohn MJ, Rachel R, Fuchs T, Wimmer VC, Stetter KO (2002) A new phylum of Archaea represented by a nanosized hyperthermophilic symbiont. *Nature* 417: 63–67.

Kajander EO, Ciftcioglu N, Miller-Hjelle MA, Hjelle JT (2001) Nanobacteria: controversial pathogens in nephrolithiasis and polycystic kidney disease. *Current Opinion in Nephrology and Hypertension* 10: 445–452.

Kasting JF, Brown LL (1998) The early atmosphere as a source of biogenic compound. In: Brack A (ed) *The Molecular Origins of Life*. Cambridge University Press, Cambridge, UK, pp 35–56.

Khanna M, Stotzky G (1992) Transformation of Bacillus subtilis by DNA bound on montmorillonite and effect of DNase on the transforming ability of bound DNA. *Applied Environmental Biology* 58: 1930–1939.

Kompanichenko VN (1996) Transition of precellular organic microsystems to a biotic state: environment and mechanism. *Nanobiology* 4: 39–45.

Koshland DE (2002) The seven pillars of life. *Science* 295: 2215–2216.

Knoll AH (1999) A new molecular window on early life. *Science* 285: 1025–1026.

Lahav N (1999) *Biogenesis: theories of life's origin*, Oxford University Press.

Lazcano A (1994) The RNA world, its predecessors and descendants. In: Bengtson S (ed) *Early life on Earth*, Columbia University Press, New York, pp 70–80.

Lorenz MG, Wackernagel W (1987) Adsorption of DNA to sand and variable degradation rates of adsorbed DNA. *Applied Environmental Microbiology* 53: 2948–2952.

Luisi PL (1998) About various definitions of life. *Orig. Life Evol. Biosphere* 28: 613–622.

Lwoff A. (1962) *Biological order.* MIT Press, Cambridge, MA.

MacKay AL (1986) The crystal abacus. In: Cairns-Smith AG, Hartman H (eds) *Clay Minerals and the Origin of Life.* Cambridge University Press, Cambridge, UK, pp 140–143.

Margulis L, Sagan C (1995) *What Is Life?* Simon & Schuster, New York.

Mason B, Berry LG (1968) *Elements of mineralogy.* W.H. Freeman and Co., San Francisco.

Maturana HR, Varela FJ (1981) Autopoiesis and cognition: the realization of the living. *Boston Studies in the Philosophy of Science* 42, D. Reidel, Boston.

McClendon JH (1999) The origin of life. *Earth Science Reviews* 47: 71–93.

McKay DS, Everett KG, Thomas-Keprta KL, Vali H, Romanek CS, Clemett SJ, Chillier XDF, Maechling CR, Zare, RN (1996) Search for past life on Mars: possible relic biogenic activity in Martian Meteorite ALH84001. *Science* 273: 924–930.

Melosh HJ (2003) Exhange of meteorites (and life?) between stellar systems. *Astrobiology* 3: 207–215.

Monnard P-A, Apel CL, Kanavarioti A, Deamer DW (2002) Influence of ionic solutes on self-assembly and polymerization processes related to early forms of life: implications for a prebiotic aqueous medium. *Astrobiology* 2: 139–152.

Monod J (1971) *Chance and necessity.* Alfred A. Knopf, New York.

Moreno A, Fernandez A, Etxeberria A (1990) Cybernetics, autopoiesis and definition of life. In: Trappl R (ed) *Cybernetics and Systems 90.* Singapur: World Scientific, pp 357–364.

Newman ACD, Brown G (1987) The chemical constitution of clays. In: Newman, ACD (ed) *Chemistry of clays and clay minerals*, John Wiley & Sons, New York, Mineralogical Society, pp 1–128.

Nielsen PE (1993) Peptide nucleic acid (PNA): a model structure for the primordial genetic material. *Orig. Life Evol. Biosphere* 23: 323–327.

Orgel LE (1998) The origin of life – a review of facts and speculations. *Trends Biochem. Sci.* 23: 491–495.

Paget E, Jocteur-Monrozoir L, Simonet P (1992) Adsorption of DNA on clay minerals: protection against DNaseI and influence on gene transfer. *FEMS Microbiology Letters* 97: 31–40.

Purves WK, Orians GH, Heller HC, Sadava D (1998) *Life: The science of biology.* Sinauer Associates, Sunderland, Massachusetts.

Random House (1987) *Dictionary*, 2nd edition, Random House, New York.

Raven PH, Johnson GB (1999) *Biology.* McGraw-Hill, Boston.

Russell MJ, Hall AJ (1997) The emergence of life from monosulfide bubbles at a submarine hydrothermal redox and pH front. *J. Geol. Soc. London* 154: 377–402.

Russell MJ, Hall AJ (2002) From geochemistry to biochemistry: chemiosmotic coupling and transition element clusters in the onset of life and photosynthesis. *Newsletter of the Geochemical Society* 113: 6–12.

Sagan C, Salpeter EE (1976) Particles, environments, and possible ecologies in the jovian atmosphere. *Astrophys. J. Suppl. Ser.* 32: 624.

Schieber J, Arnott HJ (2003) Nannobacteria as a by-product of enzyme-driven tissue decay. *Geology* 31: 717–720.

Schöning K-U, Scholz P, Guntha W, Wu X, Krishnamurthy R, Eschenmoser A (2000) Chemical etiology of nucleic acid structure: The alpha-threofuranosyl-(3′2′) oligonucleotide system. *Science* 290: 1347–1351.

Schopf JW (1994) The oldest known records of life: early Archean stromatolites, microfossils, and organic matter. In: Bengtson S (ed) *Early life of Earth.* Columbia Univ. Press, New York, pp 193–206.

Schrödinger E (1944) *What is life? the physical aspect of the living cell.* Cambridge University Press, Cambridge, UK.

Schulze-Makuch D (2002) At the crossroads between microbiology and planetology: a proposed iron cycle could sustain life in an ocean – and the ocean need not be on Earth. *ASM News* 68: 364–365.

Schulze-Makuch D, Guan H, Irwin LN, Vega E (2002) Redefining life: an ecological, thermodynamic and bioinformatic approach. In: Palyi G, Zucchi C, Caglioti L (eds) *Fundamentals of Life.* Elsevier, pp 169–179.

Schwartz AW (1993) Biology and theory: RNA and the origin of life. In: Greenberg JM, Mendoza-Gomez CX, Pirronello V (eds), *The chemistry of life's origins.* Kluwer Acad. Publ., Dordrecht, pp 323–344.

Stetter KO (1998) Hyperthermophiles and their possible role as ancestors of modern life. In Brack A (ed) *The molecular origins of life*, Cambridge University Press, Cambridge, UK, pp 315–335.

Stevens TO, McKinley JP (1995) Lithoautotrophic microbial ecosystems in deep basalt aquifers. *Science* 270: 450–454.

Szent-Györgyi A (1972) *The living state, with observations on cancer.* Academic Press, New York.

Wächtershäuser G (1994) Vitalists and virulists: a theory of self-expanding reproduction. In: Bengtson, S (ed) *Early life on Earth.* Columbia University Press, New York, pp 124–132.

3 Lessons from the History of Life

The discussion of life on other worlds is inevitably qualified by the phrase, "life as we know it." This customary and appropriate caution among scientists serves to (1) admit that all our speculations and extrapolations are based on a known sample size of only one, and (2) imply that the one form of life we know may be peculiar to the physical conditions under which it exists. While these constraints do place boundaries on the scientific latitude we should allow ourselves in speculating about unknown forms of life, the sample with which we are familiar does constitute a specific and robust example that has persisted through numerous crises in variable, changing, and often extreme environments. Assuming that the laws of physics and chemistry are universally operative, then life elsewhere might be expected to follow the same evolutionary and ecological principles that have characterized its history on Earth. Thus, rather than thinking of "life as we know it" in terms of constraints, this chapter explores the insights to be gained by regarding the one life we know as a harbinger and example of the life we can reasonably expect to exist on other worlds.

3.1 A Brief History of Life on Earth

The Earth is presumed to have formed about 4.55 billion years ago by accretion, like all concentrations of matter in the universe, through gravitational collapse and consequent rotation (Cassen and Woolum 1999). Recurrent bombardment continued for roughly 0.5 billion years, during which the Earth's interior differentiated and the atmosphere stabilized in composition, consisting primarily of N_2, CO_2, and NH_4 (Tajika and Matsui 1993).

Based on dating of the earliest fossils, the first life on Earth appeared within 0.5 billion years after sterilizing bombardment had ceased (Chyba and McDonald 1995, Schopf 1999). While evidence for the ultimate ancestry of life on Earth is controversial at this time (e.g. Mojzsis et al. 1996, Westall et al. 2001, Brasier et al. 2002), few experts doubt that life was present by 3.5 billion years ago, and some believe it is likely to be closer to 4.0 billion years old (Lazcano and Miller 1994).

The earliest life was unicellular, microscopic, and anaerobic (Schopf 1983) and this type of life still persists today. Though far more complex than any non-living particle of comparable size, its internal structure was relatively undifferentiated, lacking internal membranes or extensive subcellular segregation of function. Once the simple architecture of the prokaryotic cell emerged, it remained relatively unchanged for two billion years, or half of the history of life.

Based on fossil evidence photosynthesis arose fairly early, to harvest the abundant source of energy from the sun (Cowen 1995, Margulis and Sagan 1995). Once oxygen began to be produced by photosynthesis, it was first consumed by the oxidation of minerals, then gradually began to increase in the atmosphere. Over a billion years of photosynthesis took place before the oxygen content of the atmosphere reached 10% of its current level (Walker 1977).

Gradually, subcellular specialization developed in concert with the availability of more efficient oxidative metabolism, enabling the enlargement of cells to the macroscopic size of eukaryotes (Han and Runnegar 1992, Margulis and Sagan 1995). Life persisted, however, exclusively as unicellular and water-borne for perhaps another half billion years, and unicellular descendants of early microbial life thrive to the present day.

After the mineral capacity of the Earth to absorb oxygen became saturated, and free O_2 began to accumulate in the atmosphere, oxidative metabolism became sufficiently available to support multicellular structures. These forms of life remained small and probably sluggish for close to a billion years (Cowen 1995).

Widespread glaciation about 600 million years ago (Ma) was followed by a warming trend that coincided with an explosive diversification of structural forms and lifestyles between 500 and 600 million years ago (Cowen 1995). The diversification of animal morphology occurred quite sharply at about 540 Ma. Many of life's experimental forms did not survive an environmental crisis about 40 million years later, however. Most of the extant higher order taxa of plants and animals were fixed at that time and have remained essentially unchanged to the present.

Life emerged with difficulty from the water less than 500 Ma (Margulis and Sagan 1995). Those organisms that did so, however, quickly radiated into a variety of forms that occupied niches above, on, and beneath the terrestrial surface, which affords a higher degree of environmental heterogeneity.

Life has been recurrently inventive, with a tendency toward increasing complexity in a minority of forms over time (de Duve 1995). Most of the diversions from basic forms and strategies have become extinct relatively quickly. A few innovations have proven highly adaptive, and have persisted with little modification once they became established (Eldredge 1985). The general rule for biomass as a whole, however, has been to remain simple.

The subsurface of the ocean, and subterranean habitats even more so, provide much more stable environments than the air or the surface of the water and earth. Much if not most of the Earth's biomass has evolved to occupy these niches, where it remains relatively simple, microscopic, and unchanged in basic form and function from the early days of life on Earth.

3.2 Lessons from the History of Life on Earth

A number of generalizations are suggested by a consideration of the history of life on Earth, as outlined above. While we lack evidence at the present time that any of these generalizations apply to life on other worlds, if that life is subjected to the same physical and chemical principles that apply on Earth, there is no reason to assume that the same patterns and principles would not apply to the trajectory of life wherever it occurs. Those generalizations and their implications are elaborated below.

3.2.1 Life Arises Relatively Quickly under Conducive Conditions

The fact that life arose or took hold relatively soon after sterilizing bombardment had subsided suggests that it is not a highly improbable phenomenon in relation to the time available for it to occur. Whether life originated on Earth or was transported here from another point of origin, the one example that we have shows the origin or colonization of life to be achievable within a few hundred million years of the opportunity for it to happen (de Duve 1995, Lahav 1999), and some have argued that as little as a few thousand years may be required (Lazcano and Miller 1994).

It follows that if life could take hold on Earth within tens to hundreds of millions of years, it could do so anywhere under comparable circumstances in the universe. Since comparable circumstances (cessation of constant bombardment, presence of a liquid solvent, and reduced compounds or other forms of energy) probably existed on other planetary bodies in our solar system at one time, and surely have existed redundantly throughout the universe over the 12 billion years or more since the first galaxies formed, life itself may well have originated or existed for some time on other bodies in our solar system, and almost surely is part of the extended fabric of the universe as a whole.

The criticism can be raised that the relatively sudden establishment of life on Earth is exceptional, rather than indicative of an intrinsic tendency. However, the known window of opportunity during which life did emerge – between the end of sterilizing bombardment and the confirmed existence of life – provides a quantitative basis for estimating the probability that life would arise within a specified interval elsewhere. Assuming this window to extend from 25 to 600 million years, Lineweaver and Davis (2002) calculated a probability of 0.13 at the 95% level of confidence for the origin or establishment of life on any suitable body with an age of at least one billion years. This amounts to saying that 13 out of every hundred suitable planetary bodies with an age of at least one billion years has a significant probability of harboring life. While the number of "suitable bodies" in the universe is not known, it has to be very large, even if only a small fraction of the trillions of stars in the universe contain such planetary bodies. While statistical probability cannot prove the actuality of specific events, it does provide a quantitative argument for the plausibility of the widespread existence of life in the cosmos. Specialists in the theory of complexity, in fact, suggest that given the proper materials, appropriate physical conditions, and a flow of energy, that matter will form itself inevitably into complex, self-organizing systems (Morowitz 1968, Goodwin 1994, Kauffman 1995). Once these systems become self-propagating, the basic criteria for life have been met.

3.2.2 Life Tends to Stay Small and Simple

Size and complexity are, of course, relative terms, but in relation to the size and complexity into which some forms of life have evolved, the vast majority of the biomass on Earth, even today, is microscopically small and no more complex than the solitary eukaryotic cell.

Physiology favors simplicity, and simplicity is aided by small size. The ratio of surface-to-volume decreases inversely as size increases. The simplest living functions (physiological processes) depend critically on exchange of materials across the boundaries of the system (external membrane). Not only does the high surface-to-volume ratio of small compartments favor exchange of materials, the ability of those materials to migrate to and from the center of the cell by diffusion, the simplest mode possible, depends on having a cell radius small enough for diffusion to be a practical mechanism for movement.

Ultimately, some advantages are gained by increased complexity. Multicellular organisms can achieve greater mobility and enhanced capacity to deal with a specific range of environmental fluctuations, but multicellularity requires specializations for distribution of materials, ingestion and excretion, and coordination of different body parts. This requires greater hereditary information for coding development and physiological coordination, consumes more energy, requires more space, and draws more resources from the environment. The density of such organisms is thereby reduced. Also, while advantages accrue for adaptation to specific niches, flexibility is diminished so that overall fitness to a broad range of changing conditions over time remains with the simpler structures and functions that require less coding, smaller size, and less elaborate cellular engineering.

It is not apparent whether life intrinsically requires two billion years to evolve into multicellular complexity, or if conditions on Earth simply precluded it from happening sooner. The lack of oxidative metabolic capacity has been cited as one characteristic that possibly was limiting. The constancy of the oceans, chemically and thermally, over the prolonged dawn of the history of life may have been another. Whatever the answer, there is no reason to assume that the tendency for life to remain simple and small on

Earth would not be true of life on other worlds where conditions have long remained constant.

3.2.3 Most Organisms Remain Relatively Unchanged over their Evolutionary Life Spans

While life may arise or take hold fairly quickly, it has little propensity to evolve very fast in general. It is perhaps surprising, in fact, that given the rapid pace of which evolution is sometimes capable, little evolution at the macroscopic level took place for over half of the entire history of life on Earth. And when life did achieve multicellularity, it took further millions of years for any of it to abandon the aquatic cradle of its birth and earliest nourishment (Margulis and Sagan 1995).

Every major group of organisms consists of some members whose lineage is easily traceable to a point early in the appearance of the taxon. Salamanders, turtles, crocodiles, armadillos, sturgeons, and sharks are vertebrate examples. Clams, lobsters, cockroaches, and jellyfish are just a fraction of the invertebrate examples. Within individual species, constancy is the rule, and extinction is much more common that gradual change (Gould 1981, Eldredge 1985).

In theoretical terms, we understand the tendency for living systems to retain their basic form and function for long periods of time in the context of *stabilizing selection* (Campbell 1996, p 431). Under constant conditions, the mutations, genetic recombinations, and genetic drift that lead to deviations from the optimum phenotype are less adaptive and tend to be selected against. Those traits that are nearest the optimum are favored most strongly. So long as the environment remains stable, the optimal adaptations to it will remain the same. Outlying phenotypes will be selected against, and phenotypes conforming most closely to the optimum will be selected for, over time. The frequency distribution of traits will narrow and stabilize around the optimal, providing the highest proportion of organisms with the greatest adaptive advantage (Fig. 3.1a). Thus well adapted, the organisms will remain static as long as the environment does so.

Most of the planets and satellites in the solar system have been static for long periods of time (Buratti 1999). It seems likely that all the planets except for Venus, Earth, and Mars assumed their current form and characteristics soon after their accretion. Of the sizeable satellites, only the Moon, Io, and possibly Titan have not been ice covered for most of their existence. The barren surface of Mercury, the swirl of dense atmospheres on the gas giants, and the frozen worlds of the gas giant satellites all provide environments that have been constant for billions of years. Since the interiors of the "snowball" satellites are insulated by ice layers from the radiation, vacuum, and cold of outer space, whatever life may have gained a foothold there would be subjected to stabilizing selection in the extreme, unless changes in the flow of energy from the interior would have altered the course of physical conditions below the surface of those bodies, as might have occurred on the smaller satellites after their radiant cores expended their capacity for nuclear decay. On Venus and Mars, where irreversible changes have apparently transformed ocean-bearing planets like Earth to an oppressive greenhouse on the one hand, and a cold arid desert on the other, stable environments may still be found beneath the surface of both, and possibly in the dense cloud layers of the lower atmosphere on Venus. In short, the dynamic nature of the Earth's biosphere appears to be exceptional in the solar system. Most of the worlds close at hand provide environments of enduring constancy. If life on those worlds has followed the trajectory of life under constant conditions on Earth, we can expect that life on those bodies has been relatively unchanged from the ancient forms that characterized its beginnings.

Combining this point with the previous one above, whatever life exists on other worlds in our solar system is likely to be microscopic, relatively simple, and basically unchanged from its earliest forms. Possibly Europa, and conceivably Titan and Triton could harbor exceptions, under circumstances to be argued later. The special histories of Venus and Mars may likewise raise alternative possibilities. But to the extent that our solar system is typical of others throughout the universe, most life on other worlds is likely to be small, simple, and persistent in form and function from its early origins. Thus, while the possibility of some exceptions will be argued in subsequent sections, we concur in general with the view of Ward and Brownlee (2000) that most of the living fabric of the universe belongs in the province of microbiology.

3.2.4 Evolution is Accelerated by Environmental Changes

While the normal pace of evolution under stabilizing selection is negligible, changes in either the biotic or abiotic environment can alter the pace and direction of evolution relatively rapidly (Reznick and Ghalambor 2001). The macroevolutionary examples of greatest note include the major extinctions, followed by introduction of novel forms, at times of major environmental crisis, such as the end of the Cambrian, the Permian-Triassic boundary, and the Cretaceous-Tertiary boundary (Eldredge 1985). On a microevolutionary scale, industrial melanism is often cited as the definitive example. Cryptic coloration of moths changed over a few years in concert with changes in the color of tree trunks associated with increased soot production at the onset of the industrial revolution in England (Raven and Johnson 1999, p 409). But very rapid evolution has been induced in the laboratory for a number of systems as well (Elena and Lenski 2003).

Rapid evolutionary change illustrates the principle of *directional selection* (Campbell 1996, p 431). When an environmental feature shifts in a particular direction (say the average habitat temperature gets colder), the optimum for the phenotype that is adapted to that feature (say thickness of fur) shifts to accommodate the change, so that formerly favorable phenotypes (short

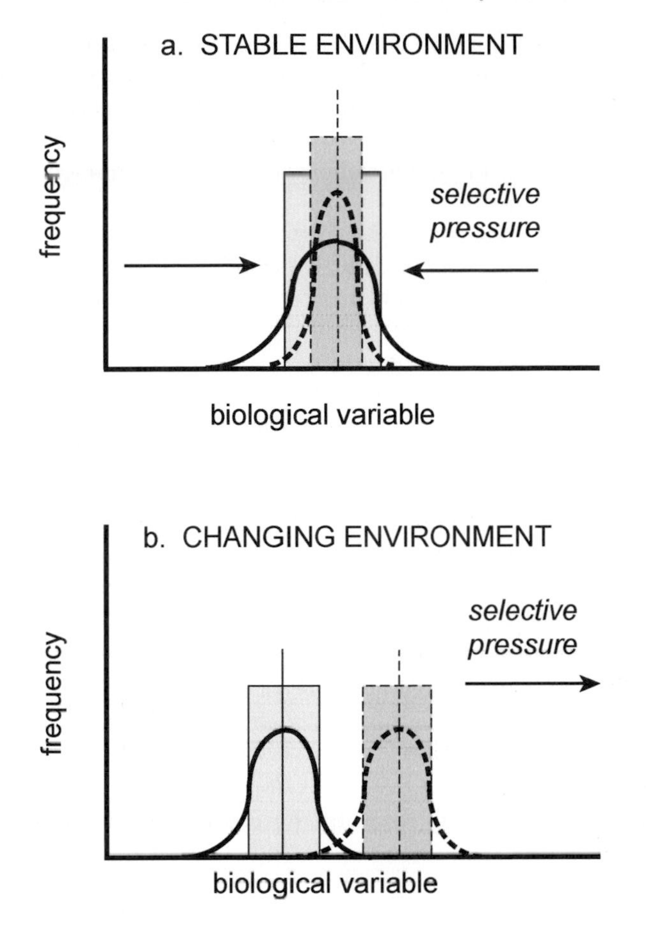

Fig. 3.1 Effect of selective pressure on biological characteristics, illustrated by changes in the frequency distribution of a quantitative biological trait in response to different forms of selective pressure. (a) In stable environments, stabilizing selection promotes elimination of peripheral values in the original population, reducing the range in the descendent population (*dashed rectangle*) without altering the mean value (*dashed vertical line*). (b) In changing environments, natural selection favors change in the direction that better adapts the organism to the new environment. The range for the majority of organisms from the original population (*solid rectangle*) and their mean value (*solid vertical line*) shift toward a different mean (*dashed vertical line*) without changing the range of the variable in the new population (*dashed rectangle*).

fur) become maladaptive and are selected against, while formerly maladaptive traits (thicker fur) become more favorable and are selected for. The frequency distribution for the phenotype, and its underlying genetic basis, shifts accordingly (Fig. 3.1b).

The more drastic and sudden the environmental change is, the quicker the evolutionary response needs to be in order for the organism to avoid extinction. In fact, extinction is more often the case, so the ones that survive are changed more radically in a briefer period of time. Over geological time spans, the fossil record thus appears to be discontinuous, with new forms arising relatively suddenly. This is the basis for the theory of "punctuated equilibrium" (Eldredge and Gould 1972, Gould 1981).

3.2.5 Complexity Inevitably Increases but as the Exception rather than the Rule

Since geological and climatic changes are very slow in relation to the life span of all organisms, from one generation to the next, stabilizing selection is the more pervasive influence, and tends to favor constancy of biological form and function. Since, as argued above, simplicity is favored over complexity by the evolutionary process, most life remains stable, and the majority of biomass remains simple. But occasional episodes of directional selection and genetic drift inevitably give rise to some forms that are better adapted by adopting a greater level of complexity.

To a degree, increased complexity equates with increased size. Thus, the evolution of the more complicated eukaryotic cell from its simpler prokaryotic ancestor was the first great leap in size and complexity in the history of life (Margulis and Sagan 1995). The evolution of multicellular organisms from unicellular ancestors was another quantum leap in complexity (Cowen 1995). Within a given size range, however, the rule of preference for maintaining the status quo continues to hold. Only a small minority of mollusks developed complex nervous systems (the cephalopods). Only a minority of vertebrates developed the complex physiological regulatory mechanisms required for homeothermy. And only a small minority of mammals developed the refined sensorimotor coordination required for manual dexterity (primates) and, eventually, complex language (humans).

But those levels of complexity did arise, and complexity theory argues in general that such complexity will arise eventually and inevitably (Gel-Mann 1994, Goodwin 1994, Kauffman 1995). This is because complexity enables specialization, and specialization sooner or later confers adaptive advantages under specialized circumstances. But specialization is difficult to reverse, as a host of interrelated organismic attributes become adapted to the specialized conditions. Once homeothermy evolves, as an example, other metabolic reactions adapt to narrow, optimal temperatures, and cease to function if the organism loses its ability to maintain the narrow thermal optimum. Thus the loss of thermal homeostasis would lead to extinction, so the specialization of homeothermy is retained as long as it provides an adaptive advantage.

Evolutionary biologists use the metaphor of a hilly landscape, where the altitude of a hill represents the degree to which its occupants are well adapted, or fit, to their particular biological niche (Dobzhansky 1951, adapted from Wright 1932). Complexity arises inevitably as time favors the "occupation" of higher points on the fitness landscape, but descent from the peaks of higher fitness is penalized by natural selection. While the overall landscape consists of fitness plains and valleys populated by less complex organisms with a broader range of general adaptations, the average level of complexity undergoes a net increase.

3.2.6 Biodiversity is Promoted by Heterogeneous Environments

Habitats can be relatively uniform over a large area (such as a desert) or volume (such as the interior of the ocean). The number of different varieties of organisms that occupy such environments is smaller than the number that occupy more diverse habitats, such as coral reefs or tropical forests (Brooks and McLennan 1991, Cowen 1995). This is because heterogeneous environments provide more niches to which different forms of life can become optimally adapted. The greater variety of living forms in turn creates more complex food webs and ecosystems. Thus, biodiversity, as measured by the overall variety of life and degree of interaction among components of the ecosystem, is greater where the habitat itself is more diverse.

By supporting a richer diversity of organisms, heterogeneous environments provide a more bountiful supply of progenitors for future evolution. Habitat fractionation thus becomes a spur to further evolution. With gradual changes in heterogeneous environments, biodiversity tends to increase even more. With sudden cataclysmic changes in such environments, biodiversity is reduced but survival of more forms is favored because of the numerical probability that more of them will be pre-adapted to persist through the cataclysm or thrive under the new environmental conditions. Thus, environmental heterogeneity not only promotes biodiversity but favors the persistence of life through challenging environmental changes.

These theoretical assumptions are well supported by the empirical evidence of life on Earth. The number of different species that occupy terrestrial habitats, which are inherently more complex than marine environments, is far greater than the number of marine species. On land, tropical rainforests provide the greatest habitat fractionation and support the greatest biodiversity. In the ocean, a much greater diversity of species is found on the continental shelves than in the deep ocean, and in the former, biodiversity is greater in coral reefs than in subtidal waters with smooth, sandy floors. The exception that proves the rule in the deep ocean is the proliferation of biodiversity around hydrothermal vents, where heated sulfur-rich effluents from encrusted "black smokers" create fragmented microhabitats with radical chemical and thermal gradients (Stetter 1985, Campbell 1996).

It follows that those other worlds in which the physical environment is heterogeneous and complex are likely to evolve a richer diversity of living forms than those with more homogeneous habitats.

3.2.7 Individuals are Fragile, but Life is Hardy

Once life evolved on Earth, it proved to be extraordinarily resilient. Despite numerous global catastrophes and recurrent environmental crises – several of which wiped out a large proportion of the species in existence – life has persisted to occupy every suitable habitat on the planet. The widespread extinctions that accompany global catastrophes (Eldredge 1985, Cowen 1995) illustrate that individual organisms lack the capacity to survive radical changes. However, at the population, species, and higher taxonomic level, the capacity for survival is more robust, because the group has a wider range of survival mechanisms than the individual.

Macroevolutionary theory is based on the view that the large-scale patterns of evolution derive from differential survival of species (Gould 1981). The ability of the group to survive despite the fragility of the individual is attributed to the concept of inclusive fitness (Raven and Johnson 1999). Contributing mechanisms include altruistic behavior, spore formation, cannibalism, fluctuating sex ratios, and adjustable reproductive strategies. At the microbial level, spores and other dormant states such as the cryptogenic state in cold environments are of special relevance, because they allow organisms to stay dormant through harsh conditions until the environment becomes suitable for survival again. An especially intriguing example is provided by *Bacillus subtilis*, a common soil bacterium. Not only can it form spores, but under starvation conditions some of the cells resist sporulation by killing sister cells, enabling them to feed on the released nutrients for survival (González-Pastor et al. 2003).

3.3 Questions Unanswered by the History of Life on Earth

While a great deal of insight can be harvested by studying historical evolution, there are some questions critical to anticipating the nature and history of life on other worlds that the history of the one life we know cannot answer.

Though life arose relatively quickly on Earth, it isn't certain that it would inevitably arise as quickly, even under the same conditions as on Earth. All we know for certain is that it *can* arise endogenously, or take hold following importation, within a half billion years of the appearance of a suitable environment.

There are no theoretical reasons or empirical observations compelling us to think that the biochemistry of life on Earth, at the molecular genetic or metabolic level, is based on the same or even similar biomolecules on other worlds. While it could be that some or all of the nucleotides, sugars, amino acids, and other metabolic intermediates that have evolved with life on Earth are so favored by physicochemical and thermodynamic selection that the same molecular configurations will inevitably evolve elsewhere, this by no means is apparent. In fact, the existence of alternative amino acids in meteorites that are not found commonly in living organisms (Cronin et al. 1988), argues against such a premise. Life on Earth shows us one form of molecular architecture that is possible, but does not rule out alternatives.

The duration of evolutionary episodes cannot be predicted from a sample size of one. While it took two billion years for eukaryotes to evolve from prokaryotes, and a billion years for macrofauna to evolve from microscopic multicellular forms, were these lengths of time necessary? The one sample we know cannot tell us whether these qualitative changes in the nature of life inherently require such long periods of time, or simply took that long for reasons that were either fortuitous, or peculiar to conditions on Earth.

The details of form and function that a different history of life would take cannot be predicted. Will taxa with calcified exteriors, such as the shells of bivalve mollusks, inevitably arise in marine environments? Will photosensitive receptors such as eyes inevitably arise if light is available? Will metabolic mechanisms for detoxifying, then utilizing oxygen for energy production, inevitably arise in the presence of oxygen? The fact that systems such as these have evolved independently under appropriate circumstances on Earth suggests the possibility that they would or at least could do so elsewhere, but such an extrapolation is not warranted by the observation of a single case.

We are left with the conclusion that life *could* arise quickly on other worlds, *possibly* using similar molecular and metabolic machinery to our own, that it *might* take as long to undergo revolutionary changes in form, size and complexity as it did on our planet, and that it *conceivably* could follow macroevolutionary trajectories that mimic the history of life on Earth. That any of these things is true cannot be ascertained from the one limiting case we have before us. It is tempting, however, to speculate that given the number of other worlds in the universe, a fraction of them could be like Earth, and that a fraction of those could have harbored life with a history similar to our own. For the vast majority of other worlds, which are dissimilar to Earth, other forms of life are clearly possible, as subsequent chapters argue. Whether these are similar to life as we know it on Earth or not, life as we do know it provides a suite of expectations about the nature of life throughout the universe, and there is no reason to doubt the validity of these general expectations.

3.4 Chapter Summary

Life arose relatively quickly on Earth, suggesting that it could do so elsewhere under appropriate conditions. Without defining (since we don't know) precisely what those conditions are, the vast number of worlds in the universe makes it virtually certain that conditions suitable for life exist on a large number of them. Therefore, life is almost surely highly redundant throughout the universe. Given the harshness and volatility of conditions at most planetary and satellite surfaces, more stable environments are to be found beneath their surfaces where the constancy of conditions favors simplicity, and stabilizing selection favors stasis of form and function. The majority of the biomass across the universe is likely, therefore, to be microscopic, simple, subsurface, and similar to its early forms. Where the physical history of the planetary body has been complex, or where the physical environment is relatively heterogeneous with a variety of boundary conditions, life can be expected to have evolved into more complex forms. In proportion to the total number of worlds harboring life, however, these are likely to constitute a distinct minority.

References

Brasier MD, Green OR, Jephcoat AP, Kleppe AK, Van Kranendonk MJ, Lindsay JF, Steele A, Grassineau NV (2002) Questioning the evidence for Earth's oldest fossils. *Nature* 416: 76–81.

Brooks DR, McLennan DA (1991) *Phylogeny, ecology, and behavior.* University of Chicago Press, Chicago.

Buratti B (1999) Outer planet icy satellites. In: Weissman M L-A, Johnson TV (eds) *Encyclopedia of the Solar System.* Academic Press, New York, pp 435–455.

Campbell NA (1996) *Biology.* Benjamin/Cummings, Menlo Park, California.

Cassen PM, Woolum DS (1999) The origin of the solar system. In: Weissman M L-A, Johnson TV (eds) *Encyclopedia of the Solar System.* Academic Press, New York, pp 35–63.

Chyba CF, McDonald GD (1995) The origin of life in the solar system: current issues. *Annu Rev Earth Planet Sci* 23: 215–249.

Cowen R (1995) *History of life.* Blackwell, Boston.

Cronin JR, Pizzarello S, Cruikshank DP (1988) Organic matter in carbonaceous chondrites, planetary satellites, asteroids and comets. In: Kerridge JF, Matthews MS (eds) *Meteorites and the Early Solar System.* Univ. of Arizona Press, Tucson, pp 819–857.

de Duve C (1995) *Cosmic dust: life as a cosmic imperative.* Basic Books, New York.

Dobzhansky T (1951) *Genetics and the origin of species.* Columbia University Press, New York.

Eldredge N (1985) *Time frames: the rethinking of Darwinian evolution and the theory of punctuated equilibrium.* Simon and Schuster, New York.

Eldredge N, Gould SJ (1972) Punctuated equilibria: an alternative to phyletic gradualism. In: Schopf TJM (ed) *Models in Paleobiology.* Freeman, Cooper, and Co., San Francisco, pp 82–115.

Elena SF, Lenski RE (2003) Evolution experiments with microorganisms: The dynamics and genetic bases of adaptation. *Nature Reviews Genetics* 4: 457–469.

Gel-Mann M (1994) *The quark and the jaguar.* W.H. Freeman & Co., New York.

González-Pastor JE, Hobbs EC, Losick R (2003) Cannibalism by sporulating bacteria. *Science* 301: 510–513.

Goodwin B (1994) *How the leopard changed its spots: the evolution of complexity.* Charles Scribner's Sons, New York.

Gould SJ (1981) G.G. Simpson, paleontology, and the modern synthesis. In: Mayr E, Provine WB (eds) *The Evolutionary Synthesis: Perspectives on the Unification of Biology.* Harvard Univ. Press, Cambridge, Massachusetts, pp 153–172.

Han T-M, Runnegar B (1992) Megascopic eukaryotic algae from the 2.1-billion-year-old Negaunee Iron-Formation, Michigan. *Science* 257: 232–235.

Kauffman SA (1995) *At home in the universe: the search for laws of self-organization and complexity.* Oxford University Press, Oxford.

Lahav N (1999) *Biogenesis: theories of life's origin.* Oxford University Press, New York.

Lazcano A, Miller SL (1994) How long did it take for life to begin and evolve to cyanobacteria? *J. Molec. Evol.* 39: 549–554.

Lineweaver CH, Davis TM (2002) Does the rapid appearance of life on Earth suggest that life is common in the universe? *Astrobiology* 2: 293–304.

Margulis L, Sagan D (1995) *What is life?* Simon & Schuster, New York.

Mojzsis SJ, Arrhenius G, McKeegan KD, Harrison TM, Nutman AP, Friend CRL (1996) Evidence for life on Earth before 3,800 million years ago. *Nature* 384: 55–59.

Morowitz HJ (1968) *Energy flow in biology.* Academic Press, New York.

Raven PH, Johnson GB (1999) *Biology.* McGraw-Hill, Boston.

Reznick DN, Ghalambor CK (2001) The population ecology of contemporary adaptations: what empirical studies reveal about the conditions that promote adaptive evolution. *Genetica* 112: 183–198.

Schopf JW (1983) Microfossils of the early Archaen apex chert: new evidence of the antiquity of life. *Science* 280: 640–646.

Schopf JW (1999) *Cradle of life: the discovery of Earth's earliest fossils.* Princeton University Press, Princeton New Jersey.

Stetter KO (1985) *Thermophilic archaebacteria occurring in submarine hydrothermal areas.* Van Nostrand Reinhold Co., New York.

Tajika E, Matsui T (1993) Degassing history and carbon-cycle of the Earth – from an Impact-induced steam atmosphere to the present atmosphere. *Lithos* 30: 267–280.

Walker JCG (1977) *Evolution of the atmosphere.* Macmillan, New York.

Ward PD, Brownlee D (2000) *Rare Earth: why complex life is uncommon in the universe.* Springer-Verlag, New York.

Westall F, de Wit MJ, Dann J, van der Gaast S, de Ronde CEJ, Gerneke D (2001) Early Archean fossil bacteria and biofilms in hydrothermally-influenced sediments from the Barberton greenstone belt, South Africa. *Precambrian Research* 106: 93–116.

Wright S (1932) The roles of mutation, inbreeding, crossbreeding, and selection in evolution. *Proceedings of the Sixth International Congress of Genetics*, pp 356–366.

4 Energy Sources and Life

An external energy source is a necessary condition for life, because living systems require a flow of energy to organize materials and maintain a low state of entropy (Morowitz 1968). Energy is also needed to perform work. Life on Earth can be distinguished by the external energy source that it uses. Autotrophic life derives energy from either sunlight or redox-reactions involving abiological compounds, while heterotrophic life uses high-energy organic molecules produced by autotrophic life as a source of energy. On other worlds, where other forms of energy may be more abundant, or where the primary sources for energy on Earth may be lacking, life may have evolved to depend on different forms of energy. In this chapter, we critically analyze the various forms of energy that are potentially available to living systems, consider other factors that bear on the evolution of energy-harvesting mechanisms, and evaluate the apparent availability of different forms of energy at different sites in our solar system.

4.1 Life As We Know It

Light, organic molecules, and oxidizable inorganic chemicals provide abundant sources of energy on Earth, so living systems have evolved specialized adaptations to use these energy sources.

4.1.1 Oxidation-Reduction Chemistry as an Energy Source for Life

Energetically favorable redox-reactions are the basis for life on Earth. The best known and most common types of metabolism are based on hydrogen oxidation and methanogenesis, sulfur reduction and oxidation, iron and manganese reduction, denitrification, and aerobic respiration. However, many other energy-yielding redox-reactions are known that involve the reduction or oxidation of relatively rare elements such as arsenic, selenium, copper, lead, and uranium. Thus, there does not appear to be a basic limitation on which elements or redox-reactions can be used. Rather, the reactions that occur will likely be dictated by the abundance, availability, and suitability of a specific element in a certain type of environment. The diversity of arsenic-

and selenium-respiring bacteria thriving in playas (alkaline salt lakes) and mining tailings is a good example (Stolz and Oremland 1999).

As pointed out in Chap. 2, the oxidation of hydrogen may be one of the most ancient and basic metabolic pathways for life on Earth, and possibly elsewhere. The oxidation of the most common element in the universe yields an appreciable amount of energy, 2.5 eV per reaction (or 237.14 kJ/mole, 56.68 kcal/mole) assuming standard conditions (25 °C, 1 atm)

$$H_2 + \frac{1}{2}O_2 \longrightarrow H_2O \tag{4.1}$$

The metabolic pathway is called methanogenesis if the oxidation of hydrogen is coupled with the reduction of carbon dioxide to methane.

$$4H_2 + CO_2 \longrightarrow CH_4 + 2H_2O \tag{4.2}$$

Methanogenesis as defined here does not imply that the hydrogen has to be supplied in molecular form, but it may also derive from an organic source. The reduction of carbon dioxide to methane requires the expenditure of energy, but due to the production of two water molecules the reaction is energy-yielding (1.4 eV energy yield at standard conditions, 474.28 kJ/mol). This reaction powers autotrophic life at hydrothermal vents and also some of the endolithic life present in the cracks and pores of the basaltic ocean floor. In addition to providing energy for metabolism, this pathway has the advantage of fixing carbon dioxide that can further be used for organic synthesis reactions. Another important redox reaction is the oxidation of molecular hydrogen coupled to the reduction of iron.

$$H_2 + 2Fe(III) \longrightarrow 2H^+ + 2Fe(II) \tag{4.3}$$

The energy yield from this reaction is 1.6 eV (148.6 kJ/mol). Examples of terrestrial organisms that use this reaction are *Pseudomonas* sp. (Balashova and Zavarzin 1980) and *Shewanella putrefaciens* (Lovley et al. 1989). There are many other compounds that can be coupled to the oxidation of hydrogen. One intriguing example is the reaction of hydrogen peroxide (H_2O_2) with molecular hydrogen to water, which is performed by *Acetobacter peroxidans* (Doelle 1969). This can serve as a model pathway for highly oxidized environments not commonly found on Earth.

Sulfur metabolism appears to be very ancient, as many terrestrial microbes are thermophilic and are associated with expressions of volcanic activity such as hot springs. One of the reasons that sulfur is so widely used is that it occurs in a wide variety of oxidation states including fractional nominal oxidation states. Oxidation states for sulfur of +7, +5, +4, $+3\frac{1}{3}$, +3, +2.5, +2, −0.4, −0.5, $-\frac{2}{3}$, −1 are known, leading to a complex inorganic sulfur cycle, much of which is mediated by microbes (Amend and Shock 2001). Sulfur-reducing bacteria are quite commonly observed to populate anoxic sediments of bottom fresh waters, and marine and hypersaline aquatic environments including submarine hydrothermal vents. Many sulfate reducers prefer molecular hydrogen or hydrogen from an organic source such as acetate or lactate, which is used as an electron donor. Some of the sulfur-reducing bacteria live symbiotically with green sulfur bacteria that photooxidize hydrogen sulfide to elemental sulfur (early photosystem). Sulfur-oxidizing microbes are common in oxygen-rich aquatic ecosystems and ground-water systems, often in close proximity to sulfur reducing microbes (Schulze-Makuch et al. 2003). Sulfur oxidation may be coupled to the reduction of iron as in the case of *Thiobacillus thiooxidans* or the thermophilic microbe *Sulfolobus acidocaldarus* (Brock and Gustafson 1976, Lovely 1991).

$$\mathrm{S^0 + 6Fe(III) + 4H_2O \longrightarrow HSO_4^- + 6Fe(II) + 7H^+} \tag{4.4}$$

This reaction yields a high amount of energy, 2.6 eV per reaction under standard conditions, and occurs in acidic environments.

Iron and manganese reduction occur in those environments associated with hydrothermal vents in the oceanic basalt crust as well. Iron reduction is coupled to the oxidation of hydrogen (see (4.3)) or the oxidation of organic sources such as glucose, lactate, formate, and organic acids. Under oxygen-rich conditions the energetically favorable reaction is the oxidation of iron, which can already occur under slightly oxidizing conditions.

$$\mathrm{Fe(II) + \frac{1}{4}O_2 + H^+ \longrightarrow Fe(III) + \frac{1}{2}H_2O} \tag{4.5}$$

This kind of reaction is performed on Earth, for example, by *Gallionella* and *Thiobacillus ferrooxidans* (note that *T. ferrooxidans* can grow by oxidizing sulfur as well as iron). The net gain of energy is 0.5 eV for each H^+ that is oxidized. Schulze-Makuch (2002) suggested an iron cycle between anaerobic ocean bottom water and oxidized upper ocean water for ice-covered planetary oceans (such as the one suspected on Europa) as a possibility for a primitive microbial ecosystem.

Denitrification and aerobic respiration are metabolic pathways that likely developed later as the Earth's atmosphere became enriched in nitrogen and oxygen. Thus, they don't relate directly to the primordial stages of life. However, denitrification and aerobic respiration are among the highest energy-yielding metabolic pathways, and thus are extremely important for the later evolution of life. Heterotrophic organisms that use pre-existing, energetically rich organic macromolecules are usually considered to have evolved later as well. However, speculation has long held that the very earliest organisms on Earth were heterotrophic rather than autotrophic (Oparin 1938, Haldane 1954, Fox and Dose 1977), based on the fact that heterotrophs have a simpler molecular machinery, and the assumption that energetically rich organic macromolecules may have been supplied on the early Earth in abundance by prebiotic synthesis (Miller and Orgel 1974) and/or by comet delivery (Chyba et al. 1990).

4.1.2 Light as an Energy Source for Life

Probably one of the most important factors in the success of life on Earth is the use of a narrow band of electromagnetic radiation (visible light) emitted from the sun at a high intensity. The evolution of photosynthesis allowed life to tap into a practically unlimited source of energy. Some time early in the history of the Earth, photosynthesis developed as a favorable energy-capture mechanism.

The free energy that can be provided by photoautotrophy can be calculated by multiplying the frequency (f) of the light that is used by Planck's constant h).

$$W = hf \tag{4.6}$$

The frequency varies inversely with wavelength, and somewhat different wavelengths are used by different organisms. For example, bacterial chlorophyll uses a wavelength of 800 to 1000 nm, while carotenoids in plants absorb wavelengths ranging from 400 to 550 nm. The average energy gain across the visible spectrum is about 2 eV (190 kJ/mol, 45 kcal/mol). It is interesting to note that this value is very similar to the free energy provided by hydrogen-oxidizing prokaryotes thought to represent early chemoautotrophic organisms. Thus, from a purely energetic viewpoint, light energy and chemical energy are equally competitive.

4.2 Life Unknown to Us

Light, organic molecules, and oxidizable inorganic chemicals are abundant sources of energy on Earth. Thus, living systems on our planet have evolved to use these forms of energy. However, living cells make biological use of other forms of energy as well. They are sensitive to heat, touch, stretch, convection of air and fluids, gravity, and pressure, as well as light and the chemicals used for energetic transformations and biosynthesis. All these stimuli generate responses through membrane transductions, either by altering the gating of ions that change transmembrane potentials, or by initiating metabolic changes through secondary messenger systems. Thus, they indicate the capacity of living systems to use energy in a variety of forms to affect biological processes (Schulze-Makuch and Irwin 2002a).

The accumulation of high-energy intermediates such as adenosine triphosphate (ATP) depends on the generation of proton gradients across membranes. Since the molecular machinery of cell membranes mediates most sensory transductions, it is reasonable to assume that natural selection could have favored the evolution of membrane mechanisms that transform heat, pressure, stress, magnetic fields, or kinetic energy into high-energy covalent bonds, either directly or indirectly by coupling to ion or proton pumps. Perhaps these mechanisms have failed to evolve in living systems on Earth only because of the lack of a need for them. On other worlds where light is unavailable, natural selection would be expected to favor the evolution of alternative energy harvesting mechanisms (Schulze-Makuch and Irwin 2002a).

How much energy is needed to power a living system? We do not know, but we can get a rough idea by calculating the amount of energy used by living organisms on Earth. First, as mentioned before, both light energy (in the visible spectrum) and chemical energy (hydrogen oxidation) provide about the same amount of energy (2 eV per photon or hydrogen molecule oxidized). Second, we know how much energy is required to assemble energy-storing compounds used on Earth. For example, a energy of about 7 kcal/mol or 0.304 eV per molecule (4.86×10^{-23} kJ or 1.16×10^{-23} kcal) is required to form the terminal phosphate bond in ATP. While the chemical storage form of energy used by another form of life would not necessarily be ATP or anything like it, the amounts of energy required for ATP phosphorylation on Earth provide a known benchmark. Given the required bonding energy of 0.304 eV, it follows that a photon of light or the oxidation of a molecule of hydrogen to water provides in theory sufficient energy to convert 6 to 8 molecules of ADP to ATP, if the conversion is close to 100% efficient. With this benchmark in mind, we can speculate whether other energy sources could be as efficient as chemical energy or light on Earth to provide sufficient energy for a living organism under a different set of environmental conditions.

4.2.1 Electromagnetic Waves (other than Visible Light)

The wavelength of light used by phototrophic organisms lies in the visible and near-infrared spectrum. The fact that these specific frequencies are used may merely reflect adaptation by terrestrial organisms to the most prevalent wavelengths of electromagnetic radiation emitted from the Sun. These wavelengths are also transmitted through our atmosphere well, making them readily available to life on the surface of our planet. On other worlds the same may be true, because most habitable solar systems would likely receive light from a G-type star like our Sun (though many habitable worlds could be associated with other types of stars such as K-type stars), and most atmospheres would absorb far less UV radiation and wavelengths below near-infrared to a great extent. However, in principle other wavelengths could be harvested. The problem with more energy-rich radiation, such as that in the ultraviolet range, is its detrimental effect on many biogenic molecules such as proteins and DNA (Rettberg and Rothschild 2002). Thus, any organism using UV or more energy-rich radiation would need some kind of protection to harvest this type of electromagnetic wavelength, or would have to be based on a very different type of biochemistry. Alternatively, organisms may be able to take up less energy-rich radiation than near-infrared wavelengths. This radiation is plentiful in the universe, as any body or substance above 0 K will radiate infrared light. A prime example is hydrothermal vents on the ocean floor of Earth that emanate infrared light at a depth where photons from the Sun cannot reach (Van Dover et al. 1994, White et al. 2002). However, many photons of infrared radiation would have to be used to harvest enough energy to be biologically useful. Gusev (2001) even suggested that organisms could use radio waves to form ATP via excitement of protons at their natural resonance frequencies through Langmuir oscillations. Even less energy would be available from this source, however.

4.2.2 Thermal Heat Capacity

Radiogenic elements decay in planetary interiors and produce heat that drives all major processes inside planetary bodies. Examples for Earth include keeping the outer core liquid, establishing a protective magnetic field, and driving plate tectonics. Tidal flexing can also cause heat to emanate as observed on Io, the volcanically most active planetary body in the solar system. The tidal flexing is caused by the changes in gravitational attraction to Jupiter and to some minor degree to the other Galilean satellites. Heat released by tidal flexing is also observed in the Earth-Moon system, but to a lesser degree.

This geothermally produced heat results in thermal gradients on which thermotrophic organisms could feed. We name these organisms "thermotrophs" in analogy to phototrophic and chemotrophic organisms that use light and chemical energy as basic energy source, respectively. A relatively straightforward possibility would be to harvest energy from the thermal gradients at hydrothermal vents. Thermotrophic life could harvest energy from the high heat capacity of water, which is about 4 kJ/kg K between 0 and 100 °C and 0 and 100 MPa pressure. If we assume a cell mass of 10^{-12} g comparable to that of microbes on Earth (Madigan et al. 2000), and further assume that one tenth of the cell mass is a vacuole of water from which the thermotrophic organism could extract energy, about 2.5×10^{6} eV would be obtained from cooling the vacuole by 1 °C. If a microbe were able to use the Carnot cycle, the organism could extract about 9000 eV of usable energy for a temperature change from 5 °C to 4 °C (Schulze-Makuch and Irwin 2002a). This amount of energy is much greater than the amount that can be extracted from photoautrophy (per photon) or chemoautotrophy (per molecule of hydrogen oxidized to water). For a cell as large as the giant pantropical alga, *Valonia macrophysa* (Shihira-Ishikawa and Nawata 1992), containing a water vacuole of approximately 10 g, the potential energy yield could be close to 1 Joule. Also, energy could be harvested from a temperature differential much greater than 1 °C.

Temperature gradients that may exist between the vacuole and the surrounding cell material can only be harvested when heat flows from a hot object to a cooler one. A "thermotroph" could heat up the water vacuole by having it located very close to a hydrothermal vent. It could then detach from the substrate, float in the colder ocean and harvest the heat flow between the vacuole and the surrounding cell plasma (Fig. 4.1). The vacuole filled with hot water would be the internal heat engine on which the organism thrives as well as giving it the buoyancy to float in the ocean water. Or, another solution would be for the "thermotroph" to be anchored at the hot discharge zone of a hydrothermal vent while its distal end extends into the colder adjacent water, similar to the elongated pogonophoran tubeworms known to populate black smokers on Earth's ocean floor. A plausible transduction mechanism could involve the presence of membrane macromolecules that catalyze high-energy metabolites through temperature-dependent conformational changes (Fig. 4.2).

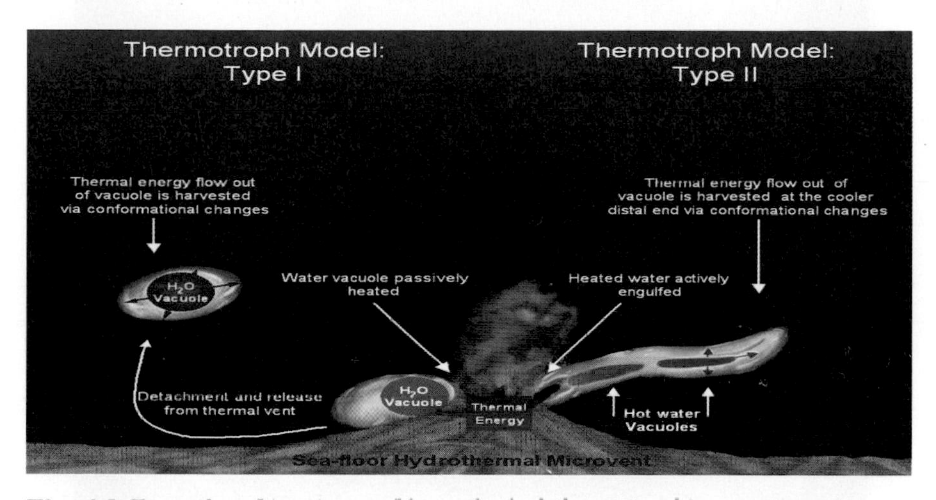

Fig. 4.1 Examples of two types of hypothetical thermotrophic organisms that use hydrothermal vent water to harvest energy. Art provided by Chris D'Arcy, Dragon Wine Illustrations, El Paso, Texas.

A potential drawback to the use of thermal energy is its inefficiency. The most efficient thermodynamic system known – the Carnot engine – is very inefficient, especially for small temperature differences. Because of the low efficiency, most of the energy in a thermal gradient would be dissipated as heat without being captured by chemical bonds, and would readily degrade the thermal gradient itself. A possible solution to the problem would be a thermotrophic organism that could shuttle back and forth across fairly sharp environmental gradients, such as that between a hydrothermal vent and surrounding cold ocean water. Alternatively, an elongated thermotrophic organism could make use of convection to dissipate the unusable entropy-related energy (as consequence of the second law of thermodynamics; see also discussion in Sect. 4.3).

Another way of harvesting energy from temperature gradients was suggested by Muller (1985). He envisioned a mechanism by which biomembranes convert heat into electrical energy during thermal cycling. This process of

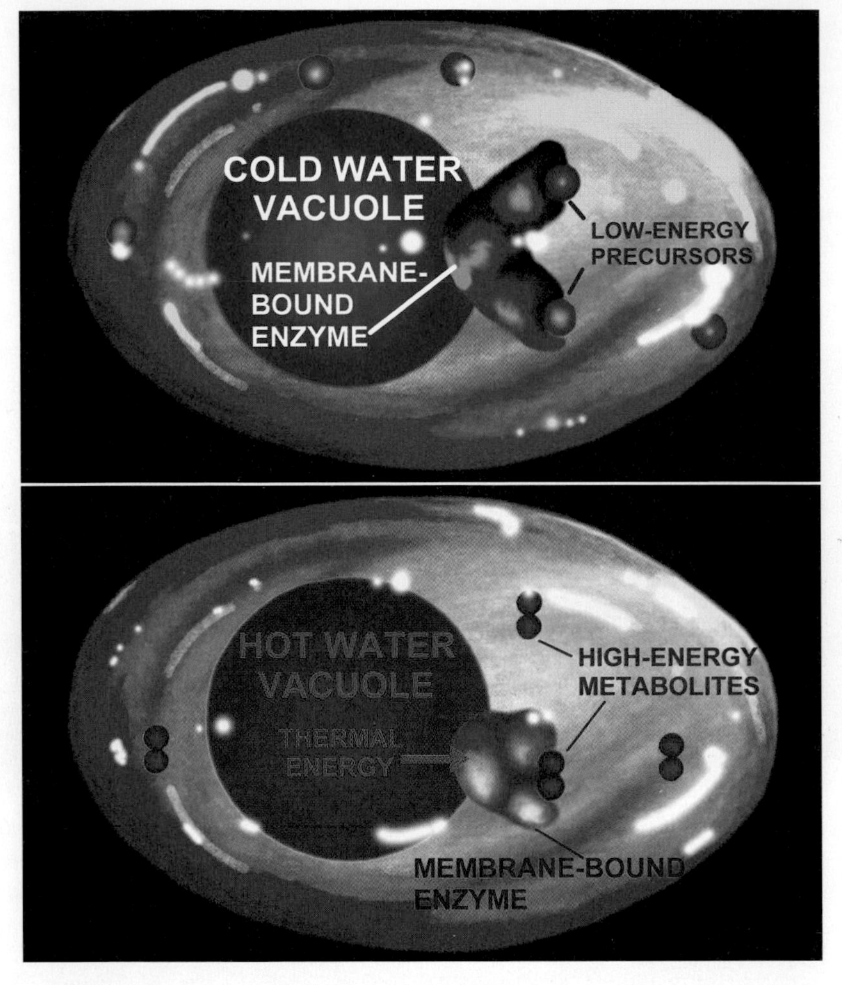

Fig. 4.2 High-energy metabolites can be produced via conformational changes if a temperature gradient between vacuole and cell plasma is present (*lower image*), but cannot form without a temperature gradient (*upper image*). Art provided by Chris D'Arcy, Dragon Wine Illustrations, El Paso, Texas.

thermosynthesis could possibly occur in leaves during cyclic transpiration and in organisms living in convecting volcanic hot springs. Electrogenic ATP synthases then could convert the electrical energy gained by thermosynthesis into ATP if their activity and stoichiometry were properly regulated. If correct, thermosynthesis could be a plausible basic pathway of metabolism for early organisms on Earth, possibly a progenitor of bacterial photosynthesis (Muller 1995, 2003), and an option for possible life on other worlds.

4.2.3 Kinetic Energy

The kinetic energy of convection cells or tidal currents could be harvested directly and used to sustain life. Organisms could contain pili or cilia much like ciliated bacteria or protozoa and adhere to a substrate at the ocean bottom or on the underside of an ice ceiling, where they are exposed to currents of moving water that can bend their cilia (Fig. 4.3). The cells could enclose protein-like macromolecules that induce an electrical polarity across the membrane through a Donnan equilibrium. The hair cells could be surrounded by Na^+ channels whose permeability is proportional to the deflection of the hairs, with properties like those of sensory hair cells in the vestibular membrane of vertebrates or lateral line organs of fish (Fig. 4.4).

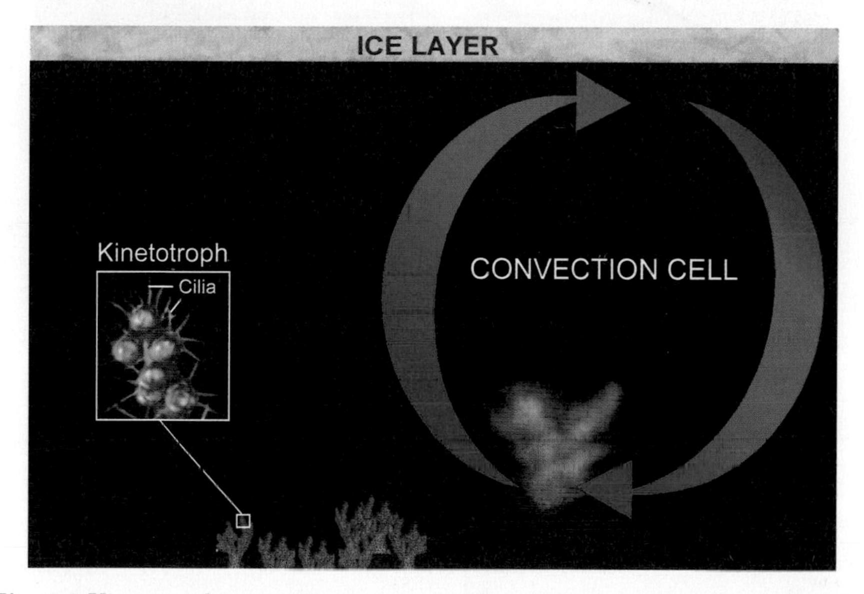

Fig. 4.3 Kinetotrophic organism in an oceanic environment, schematic. Convection currents can bend cilia leading to the opening of Na^+ channels, allowing Na^+ to flow into the cell passively down its concentration gradient. Art provided by Chris D'Arcy, Dragon Wine Illustrations, El Paso, Texas.

By bending the cilia, the convection currents could open the ion channels, allowing ions to flow into the cell passively down their concentration gradients. This thermodynamically favored process could be coupled to the direct formation of high-energy phosphate bonds or to a H^+ transporter across another internal membrane, by analogy with mitochondrial membranes (Schulze-Makuch and Irwin 2001). The ionic gradient would be maintained by extrusion of the ions via exocytosis. The ions could bind, for example, to intracellular macromolecules whose tendency to fuse with the external membrane and disgorge their ionic ligands is thermodynamically favored, once the number of ions bound to the carrier reaches a concentration greater than the concentration of the ions outside the cell (Schulze-Makuch and Irwin 2002a). Alternatively, the ions could simply be precipitated as salts. A steady convection current with a velocity in the mm/s range would certainly be able to provide the requisite molecular distortion. Since this system works essentially like a battery that is charged over time, all that is needed is a minimal ionic gradient and enough time to charge the system high enough to form energy-storing chemical compounds.

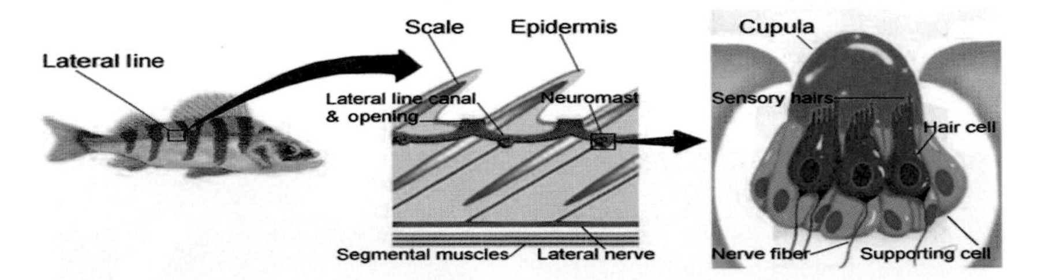

Fig. 4.4 Lateral line organ in fish, illustrating the role of cilia in transducing fluid movement into electrogenic signals. Cilia bend from the movement of fluid, allowing ions to move through the open channels. Art provided by Chris D'Arcy, Dragon Wine Illustrations, El Paso, Texas.

4.2.4 Osmotic or Ionic Gradients

Osmotic gradients can be an enormously powerful source of energy. The osmotic pressure can be calculated by the van't Hoff formula.

$$\prod = cRT \tag{4.7}$$

where $\prod$ is osmotic pressure (atm), c is the molar solute concentration (mol/L), R is the universal gas constant (0.08206 L atm/mol K), and T is the absolute temperature (K). Some halophilic strains of cyanobacteria are known to tolerate salt concentrations of up to 2.7 M NaCl (Hagemann et al. 1999). Marine teleosts (bony fishes) retain a strong osmotic differential of roughly 0.7 osmoles between their intercellular fluids and their surrounding environment (Wilmer et al. 2000), where 1 osmole is one mole of osmotically active particles. Using this conservative figure as a first estimate at a temperature of 25 °C (298 K), the osmotic pressure would be 16.9 atm (1.7×10^6 Pa). The force that acts on one water molecule along its concentration gradient is then

$$F = \prod A \tag{4.8}$$

where A is the cross-sectional area of one water molecule. This force is about 10^{-13} N. Further, assuming this force moves the water molecule through a membrane channel that couples the movement to formation of a high-energy covalent bond, the energy available for bond formation is given by

$$W = Fs \tag{4.9}$$

where s is the distance the water molecule moves down its density gradient (assumed to be 10^{-8} m for a biomembrane). Using the above figures, the calculated potential energy yield is 10^{-21} J, or 0.007 eV. Thus, one ATP could be phosphorylated from ADP for about every 45 water molecules entering the cell by osmosis. This is about two orders of magnitude below the energy yield for chemoautotrophs or photoautotrophs on Earth. The 0.007 eV may be a conservative estimate, because the osmotic differential calculated here is based on those of fish that have adapted from their freshwater origin to their marine environment rather than microbes adapted to use osmotic gradients. Halophilic microbes as described above, if adapted not only to tolerate but to use osmotic gradients, might easily be able to more than quadruple this energy yield.

The direct coupling of water movement to phosphorylation reactions is not known for living systems on Earth. However, evolution could have favored the origin of membranes in which water movement yields energy, where osmotic gradients were readily available and other forms of energy were not. A plausible mechanism would involve tertiary structural changes in a channel-associated protein that catalyzes formation of high energy bonds, much as ligand-induced conformational changes in membrane receptors lead to a series of steps culminating in the synthesis of high-energy cyclic AMP (Schulze-Makuch and Irwin 2002a). As in the case of thermal gradients, degradation of the osmotic gradient is a potential drawback to their use for generating free energy. The influx of many water molecules would either significantly increase the cell volume, or increase counteracting pressure in rigid cells that cannot expand in volume. This could be mitigated, however, by a compensatory loss of solutes, such as efflux of Na^+ (and Cl^- for electrical balance) powered by the rise in intracellular pressure. Either the extrusion of solutes, or the pressure itself, could be coupled to conformational changes that could catalyze high energy bond formation. Alternatively, cell volume could be reduced by reverse osmosis upon movement of the organism to a more hypertonic level of the liquid medium. The zone close to the ocean bottom would be expected to be high in total dissolved solids due to persistent dissolution of the mantle and a higher density of salt water compared to fresh water. Solute gradients would be present both at the ocean bottom and in higher regions, but in opposite directions if not much water circulation is occurring. For example, a membrane water channel could be coupled to a reaction that forms a high-energy bond inside the cell as the water moves inward from hypotonic surroundings, while a similar channel oriented in the opposite direction could harvest energy when water leaves the cell in hypertonic surroundings. The hypothetical organism could thus move between two layers of different salinity, using both to harvest energy (Fig. 4.5).

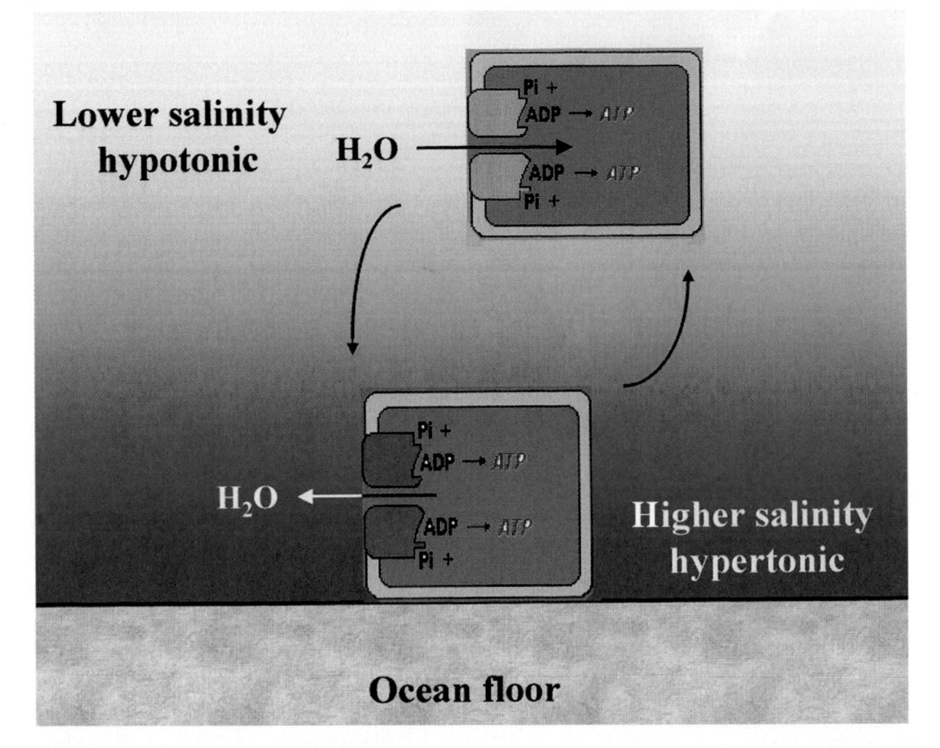

Fig. 4.5 A hypothetical osmotrophic organism that harvests energy from salinity gradients. Movement of water would be coupled to a reaction that forms a high-energy covalent bond through variants of a membrane molecular complex that are energized by entrance or exit of water, depending on the direction of the osmotic gradient (P_i = inorganic phosphate, ATP formation is used as an example, not necessarily implying that ATP is used by a hypothetical osmotrophic organism; from Schulze-Makuch and Irwin 2002a).

Ionic gradients and H^+ gradients conceivably could also provide energy. For a 100-fold ionic gradient between the cytoplasm of an organism and its external environment – a differential observed in halobacteria on Earth (Madigan et al. 2000) – the amount of potential energy can be calculated from the Nernst equation

$$E = (RT/nF)\ln([\text{ion}]_{\text{ext}}/[\text{ion}]_{\text{int}}) \tag{4.10}$$

where n equals the number of charges transferred in the reaction, F the Faraday constant (J/mV mol) and R and T as above. The potential energy yield, $\Delta G = -nFE$, equals 0.12 eV, when $[\text{ion}]_{\text{ext}} = 100 \times [\text{ion}]_{\text{int}}$ at 298 K. This could drive the diffusion of about 3 ions, which could provide the energy for the phosphorylation of one ATP molecule. This might well be an underestimate of the potential energy yield since some bacteria achieve ionic distribution ratios as high as 10^6 across their membranes (Neidhardt et al. 1990).

Thus, the harvesting of osmotic or ionic gradient for bioenergetic purposes appears to be feasible in principle. Both types of gradient would often coexist, allowing for the possibility of reciprocal cycling between the two. On other planetary bodies where strong chemical concentration gradients are likely present, as in the putative liquid ocean on Europa (Kargel et al. 2000) and possibly other icy satellites, the use of osmotic or ionic gradients as bioenergetic sources must be considered a reasonable possibility, especially where other energy-yielding strategies may not be feasible. Irwin and Schulze-Makuch (2003) have recently modeled a putative multilevel ecosystem, based on the assumption of a hypertonic ocean bottom and a hypotonic ocean ceiling on Europa. Their calculations indicate that organisms the size of brine shrimp could be supported at a density of several hundred per cubic meter at the ocean bottom. While such an ecosystem is purely hypothetical at this point, their calculations point to the theoretical feasibility of an ecosystem in which the producer level is powered purely by osmotic or ionic gradients.

4.2.5 Magnetic Fields as Energy Source for Life

Charge separation and extractable free energy can be generated from magnetic fields in theory. Magnetic fields can yield energy based on the Lorentz force, the movement of a charge within a magnetic field, or by induction from a periodically changing magnetic field. The Lorentz force can be expressed by

$$F_{\text{L}} = q(E + v \times B) \tag{4.11}$$

where E is the electric field acting on the charge (Newton/Coulomb; N/C), v is the velocity (m/s) of the charge in the magnetic field, and B (T) is the magnetic field strength. The cross-product $v \times B$ is reduced to vB in the special case of a perpendicular movement of the charge with respect to the direction of the magnetic field B. If the movement of the charge occurs parallel to the direction of the magnetic field, the cross-product is zero, and thus in the absence of an electric field no force acts on the charge q. A somewhat analogous directional dependence is observed for induction. In the absence of a magnetic field ($B = 0$), a charge is accelerated parallel to the electric field such that

$$F = qE \tag{4.12}$$

with E being the magnitude of the electrical field (N/C).

4.2.5.1 Possible Biogenic Use of the Lorentz Force

The amount of energy that can be extracted via the Lorentz force depends on the strength of the magnetic field of the particular planetary body. The strength of Earth's magnetic field at the surface is about 0.3 gauss, or 3×10^{-5} Tesla (T). There are planetary bodies that have a far larger magnetic field strength than Earth, such as Jupiter (4.3 gauss at the equator) and Saturn. Earth's magnetic field can be described in a first approximation as a magnetic dipole. Any charged particle moving in a conducting liquid at a direction perpendicular to the magnetic field line would experience the Lorentz force. Protons inside organisms are charged and Earth's oceans consist of salt water, which is a conducting liquid. But would the Lorentz force be sufficiently strong to yield a significant amount of energy? Let us assume a hypothetical organism the size of a terrestrial microbe that moves within a convection cell at a rate of 1 m/s perpendicular to the magnetic field lines. Then the potential energy yield can be calculated from

$$W = F_{\mathrm{L}} s \tag{4.13}$$

where W is the energy yield (J or eV, F_{L} is the Lorentz force as given above, and s is the distance where charge separation can occur. If it is assumed that the total length of the microbe (10^{-6} m) can be used for charge separation, then the Lorentz force that can act on a unit charge (electron or proton, q =1.6×10^{-19} C) is about 5×10^{-24} N. This charge separation can be imposed if the magnetic field is perpendicular to the line of movement of the charge and be released if the magnetic field is oriented parallel to the movement of the charge. It follows that the energy obtained from the Lorentz force is 3×10^{-11} eV, 11 orders of magnitude lower than the energy that could be obtained by chemoautotrophy (assuming hydrogen-oxidizing metabolism) or photoautotrophy. However, there is no conceptual reason why charge separation of only one electron or proton can be harvested at one time; several or thousands of reactions could occur in parallel. Also, the available energy yield could be increased by many orders of magnitude, if the microbe is an analog to hair cells on Earth with lengths in the millimeter or centimeter range instead of being ball- or pancake-shaped in the micrometer range. A plausible energy-harvesting mechanism could be that H^+-ions are driven across a one-way channel against their concentration gradient into an internal organelle, where they accumulate to a higher concentration gradient than on the outside of the organelle. Then, the H^+-ions can diffuse back out of the organelle through a different channel coupled to a phosphorylation reaction that produces a high-energy organic phosphate (similar to the way mitochondria work). A more elaborate scheme would be if the Lorentz force is used to separate one electron and one proton from the center of the cell in opposite directions toward the respective ends of the hair cell. When the magnetic field lines are oriented parallel to the hair cell, the energy could be released and the proton and electron would move toward each other producing H_2 (Fig. 4.6). The efficiency of both types of model organism could be increased if its cells contain magnetite crystals that would orient themselves to the external magnetic field to optimize the harvest of magnetic energy (Schulze-Makuch and Irwin 2001).

In addition to the model organisms discussed, there is another interesting possibility: charge separation would not necessarily have to occur within the microbial cell. Organisms of microbial dimensions could be envisioned that would cling to strips of inanimate conducting material and harvest energy from the magnetically induced electron flow in their substrates. That possibility would depend, in part, on whether a suitable mineral or aggregate of conducting matter would be present under the environmental conditions in question. Mineral assemblages including silicates, zeolites, albite, sphene, and illites, plus the iron minerals pyrite and hematite, have been identified at deep hydrothermal systems on Earth (Gonzalez-Partida et al. 2000) and could constitute suitable materials. Although not impossible, it would take a considerable effort to overcome a difference of 11 orders of magnitude in energy gain. Thus, it is not surprising that we don't encounter magnetotrophic life on Earth. However, sensitivity to magnetic fields, as in magnetotactic bacteria, is a well-established phenomenon (Frankel et al. 1979, Blakemore and Frankel 1981), and there is strong evidence that some animals use magnetospheric orientation for navigation (Ioale et al. 2000, Aekesson et al. 2001). This implies the presence of cellular mechanisms for detecting magnetospheric energy. In the absence of more concentrated or effective sources of free energy, it seems plausible to assume that a cellular mechanism for transducing magnetospheric energy into a biologically useful form could evolve.

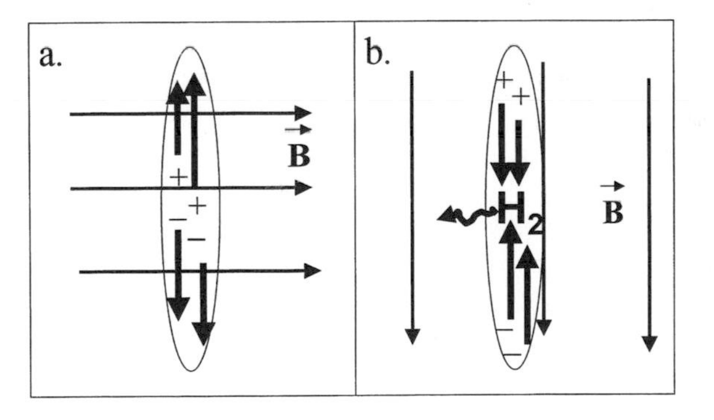

Fig. 4.6 Scheme for magnetotrophic organism to obtain energy, (a) Lorentz force separates protons and electrons, (b) magnetic field lines are oriented parallel to long axis of microbe and protons and electrons form molecular hydrogen.

4.2.5.2 Possible Biogenic Use of Induction

The second physical option for harvesting energy from a magnetic field is by induction. The possibilities and problems associated with that option will be examined using Jupiter's moon Europa as an example (Schulze-Makuch and Irwin 2002a). It now seems highly likely that Europa has a liquid ocean beneath its icy surface (Kivelson et al. 2000), which may be a suitable environment for microbial life. Jupiter's magnetospheric plasma corotates with Jupiter at a corotational velocity of 118 km/s at the orbit of Europa (Beatty and Chaikin 1990). Thus, Europa moves with respect to Jupiter's rotating field lines at a relative velocity of 102 km/s. However, the force exerted by this magnetic field cannot be used by an organism in the ocean. Europa's thick insulating layer of ice concentrates induced charges that produce an electric field that exactly cancels the Lorentz force. However, it was observed that Jupiter's magnetic field creates charge separation in a global conducting layer, which was interpreted as a shell of a salty ocean with a high electrical conductivity (Zimmer et al. 2000). Thus, the option remains that energy can be obtained from magnetic fields based on the induction from a periodically changing magnetic field (the Lorentz force, however, would still apply, as a charge can move perpendicular to the induced magnetic field, but it would be at a much lower magnitude). The alternating magnetic field that is experienced by Europa's ocean can be described as follows: Europa is subject to an oscillatory magnetic field

$$B = B_0 \sin(wt) \tag{4.14}$$

with an amplitude B_0 of about 200 nT. Thus, the rate of change is

$$\mathrm{d}B/\mathrm{d}t = wB_0 \cos(wt) \tag{4.15}$$

with a maximum value of $2\pi\ B_0/10$ hrs, which corresponds approximately to 2 nT/min (Khurana et al. 1998). The work W performed on a charge q (e.g. an electron or proton) is then given by

$$W = U_{\mathrm{ind}} q \tag{4.16}$$

where $U_{\mathrm{ind}} = A(\mathrm{d}B/\mathrm{d}t)$ with A being the microbial cross-sectional area [m^2], $\mathrm{d}B/\mathrm{d}t$ the change of the magnetic field strength [T/s], W is work or energy [J], and q is a unit charge of 1.602×10^{-19} C per electron or proton.

Assuming a microbial diameter of 1 μm, and the above figure for a reasonable maximum temporal magnetic field change in Europa's ocean, the amount of energy per reaction that can be extracted via induction is about 4×10^{-42} J or 3×10^{-23} eV per electron, which is 23 orders of magnitude lower than the energy that can be harvested via chemosynthesis or photosynthesis on Earth. Even if many of those reactions would occur simultaneously, the energy gain is much too low. Thus, induction does not appear to be a feasible option for living systems to capture energy in Europa's ocean, and would be very unlikely anywhere else.

4.2.5.3 Concluding Remarks on the Biogenic Use of Magnetic Energy

Life based on magnetic energy does not appear to be very promising in an Earth-type environment. However, there are certain possibilities if magnetic field strengths are much larger than on Earth. The gas giants Jupiter and Saturn have much larger magnetic field strengths, and neutron stars have magnetic field strengths millions of times stronger than the Sun or planets that surround a star like our Sun. Although life on gas giants or a neutron star itself would be very unlikely due to other considerations that are discussed in the next chapters, some moons that orbit around gas giants and especially planets that orbit neutron stars may provide an opportunity for organisms to harvest magnetic energy. Refined adaptation strategies of organisms may increase the efficiency of magnetic energy and make it competitive with light and chemical energy, even under planetary conditions similar to our solar system; but the strategies would have to be highly refined in a more efficient direction to make up for the generally low energy yield.

4.2.6 Gravitational Forces

Gravitational energy could be harvested in a simplistic fashion by lifting up protons or molecules against the gravitational attraction of a planetary body. The effects of tidal interactions are shown on Earth by the tidal amplitudes in the oceans caused by both Moon and Sun, and also on Io, the volcanically most active planetary body in our solar system, which is exposed to strong tidal interactions between Io on one side and Jupiter and its other three major moons on the other side. The free energy that can possibly be harvested can be calculated when assuming a simple model in which a proton is moved a micrometer (assumed microbial diameter) against the gravitational attraction of Earth

$$W = m_{\mathrm{H}^+} g h \tag{4.17}$$

The energy is about 10^{-13} eV and thus much smaller than the energy than can be harvested via chemoautotrophy and photoautotrophy. If large macromolecules instead of protons would be moved ($\sim 10^6$ a.m.u.), the energy yield could be as high as 10^{-7} eV. Even when allowing a planetary body the size of Jupiter, chemoautotrophy and photoautotrophy would still out-compete gravitational energy. Thus, gravitational fluctuations seem unlikely as a useful basis for bioenergetics for rocky planets, because gravitational forces are too weak. However, gravitational forces are the cause of convection currents that could be used to power a living system on terrestrial bodies as discussed before (see Sect. 4.2.3). It should also be pointed out that organisms are sensitive to gravity on Earth. Even single-cell organisms can orient themselves in a gravitational field through the use of intracellular receptors located in the cell membrane (Bräucker et al. 2002).

4.2.7 Tectonic Stress

Meteorite impacts and fractures within a planetary crust release energy that could be used by organisms for obtaining energy. Meteorite impacts are common occurrences as can easily be observed on our Moon, Mercury, and Mars. The meteorite's kinetic energy is converted upon impact into electrical potential, while the mechanical disruption of the impact causes the release of stress energy in the form of light, heat, electrical fields, and magnetic fields (Borucki et al. 2002). While the impact only occurs in a short time period that can be measured in milliseconds, melted slurry pools under impact sites can persist for as much as 1 million years, depending on the size of the impact, as shown by Thompson and Sagan (1992) for Titan. Additional energy supplied from crustal stress to the subsurface regions after the impact could increase the lifetime of the melted impact zone and provide suitable conditions for microbial life. While the energy released from the immediate impact of the meteorite would dissipate relatively fast, energy provided by the piezoelectric effect could be provided for a very long time. Piezoelectric energy is a form of electricity generated when a pressure is applied to an ionic solid as a result of strain. A charge across the crystal is produced because a dipole moment is created by the deformation of the otherwise non-polar molecular structure. The polarization of the crystal faces parallel to the direction of strain converts the crystal into a capacitor, which temporarily stores an electrical charge. A good example for demonstrating this effect is a planar molecule of any ionic solid. Its structure has 3 electric dipoles at 120 degrees to each other, which cancel and give a net dipole moment equal to 0. Applying a pressure or a voltage to this molecule will result in deformation of the molecule, and the generation of a dipole moment as a result of this deformation. The piezoelectric effect will immediately cease if the pressure or current is removed from the structure of the crystal. Piezoelectric energy, although somewhat "exotic", could thus provide a possible avenue for life. It could be associated not only with meteorite impacts but also with other geological events such as plate tectonics on Earth. However, it is restricted to 20 of the 32 crystal classes, ionic crystalline solids that lack a center of inversion such as quartz, perovskite, sodium chlorate, and Rochelle salt. As with the other alternative energy sources, no indication of its energetic use by microbial organisms has been observed on Earth. It is doubtful whether this process can provide a sufficient or sufficiently constant amount of energy. At the very least, it does not appear to be competitive with chemical or light energy.

4.2.8 Pressure Gradients

Life based on energy harvested from a pressure gradient is another theoretical possibility of an energy source driving metabolic reactions in microbial organisms. There are three principal opportunities. Energy could be harvested from atmospheric, fluid, or subterranian pressure gradients.

Pressure gradients exist in the vertical column of any atmosphere held by gravity. However, the capacity of such a pressure gradient to be used as an autotrophic energy source is questionable. Atmospheric pressure is the sum of the forces of all the molecules striking a surface area, and thus a measure of the linear momentum of the gas molecules. A pressure gradient is established if the molecules do not move randomly, but in a preferred direction. It is difficult to see how a microbial organism suspended in the atmosphere could utilize the pressure gradient. The organism would just be carried along with the wind, unable to maintain a consistent orientation within the pressure gradient, which in any event would be miniscule over the linear dimensions of the organism. This does not exclude the atmosphere of planets and large moons as habitable environments, but other metabolic strategies such as chemoautotrophy or photoautotrophy would have to be employed, as suggested originally for the Jovian atmosphere (Sagan and Salpeter 1976) and more recently for the Venusian atmosphere (Grinspoon 1997, Schulze-Makuch and Irwin 2002b).

Large pressure gradients can also exist in fluids. For an organism suspended in the fluid, the same problem would be faced as that of an organism suspended in an atmosphere. If the organism were attached to a fixed substrate, it could at least maintain a consistent orientation within a localized pressure gradient. Such localized pressure gradients exist, for example, at hydrothermal vents on the ocean floor. However, the wide fluctuations in pressure likely to arise at the vents would represent a practical problem. Furthermore, how an organism could stay attached to a substrate at these pressures and at the same time harvest the pressure gradient is unclear.

Tremendous pressures are present in the vertical rock column of a planetary body's lithosphere. Since microbial organisms are known on Earth to live at considerable subterranean depths, there is precedence for assuming that life on other worlds would occupy this habitat. Although the absolute pressure is high, the pressure gradient is not, especially with respect to microbial dimensions. Thus, life based on pressure gradients in the subsurface does not appear likely.

4.2.9 Spin Configurations

Atoms in molecules can revolve and rotate in various ways. For example, the two atoms of hydrogen gas, H_2, perform a vibrational motion in the direction of the line joining the nuclei, and a rotational motion around a direction perpendicular to the molecular axis in addition to the translational motion. The vibrational and rotational motions are in quantum states resulting in two different modifications of hydrogen: parahydrogen molecules that have antiparallel nuclear spins and even rotational quantum numbers, and orthohydrogen molecules that have parallel nuclear spins and odd rotational quantum numbers (Fig. 4.7). The two sets of molecules do not easily convert into each other but can be considered as two gases differing from each other in certain optical and thermal properties (Farkas 1935). The o-hydrogen is the higher energetic state but is relatively stable kinetically. For example, although at 20 K equilibrium hydrogen consists practically of pure parahydrogen, simply cooling the hydrogen to this temperature or transitory liquefaction or solidification does not cause equilibrium to be established.

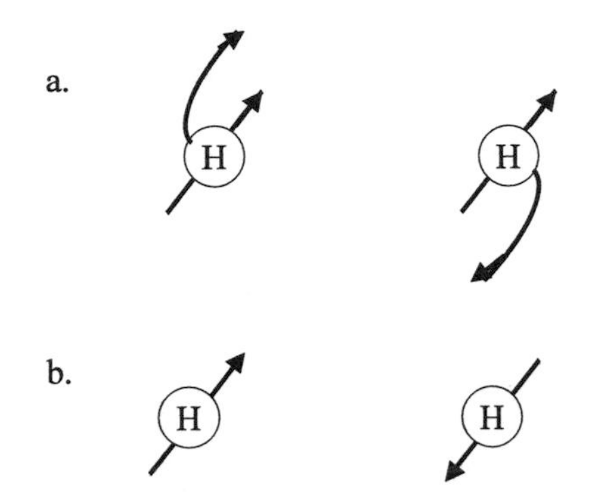

Fig. 4.7 Motions of o-hydrogen and p-hydrogen, (a) o-hydrogen atoms revolve around a common center, while their nuclei rotate around their own axis at the same time, (b) in the simplest version of p-hydrogen the atoms do not revolve and the nuclei rotate in opposite directions (modified from Feinberg and Shapiro 1980).

The two states of hydrogen could provide a source of energy in very cold environments with abundant hydrogen. For example, one could envision a mechanism to retrieve energy by having o-hydrogens on a cell boundary collide with o-hydrogen molecules in the environment. Both o-hydrogens would convert to p-hydrogen and a relative high energy yield of about 700 J/g would be obtained. P-hydrogens could then be converted back to o-hydrogens by allowing rotating p-hydrogens to collide with magnetic impurity molecules (such as oxygen), which would catalyze the formation of o-hydrogens. The potential of spin configurations as an energy source was first realized by Feinberg and Shapiro (1980), who suggested the possibility of life based on spin configurations on a very cold and dark planet, just a few tens of degrees above absolute zero. Although energy based on spin configurations represents an intriguing idea, it is doubtful that at these low temperatures energy could be transferred into chemical energy usable to organisms at a high enough rate. The idea may rather present an interesting engineering opportunity and challenge to obtain energy for space probes on cold and dark planetary bodies.

4.2.10 Radioactivity

Radioactivity is one of the most basic processes in our universe. Radioactive nuclei of atoms decay with time and release particles and electromagnetic radiations. Forms of high-energy radiation include alpha (helium nuclei) and beta particles (electrons or positrons), gamma rays (short wavelength, high-energy photons), X-rays, neutrons, and heavy ions. This type of radiation is very destructive to life as we know it because it destroys biologically important molecules, especially the sensitive machinery for molecular replication. Biological effects depend on the spatial density of ionizations produced per unit absorbed dose in the irradiated tissue. For example, particles with high atomic numbers and high energy (HZE particles) cause the greatest damage for a given dose (Baumstark-Khan and Facius 2002). Nuclear processes are the ultimate source of energy for life on Earth, since the Sun emits photons as a byproduct of the fusion of hydrogen into helium at its core. And heating produced by radioactive decay in the Earth's core could serve as the energy source for possible thermoautotrophic organisms. The question remains, however, whether radioactivity can be tapped directly by living organisms. It may be too destructive for organic synthesis and reproduction. Due to the presence of many radioactive isotopes with short half-lives, radioactive material was much more common on the early Earth, when life originated or first became established. Yet, today high-energy radiation does not serve as an energy source for any form of life as far as we know. Gamma radiation and X-rays may have been too rich in energy and too difficult to control. Or, was the visible light of our Sun simply more accessible and easier to use? How about the alpha and beta particles? They possess ample amounts of kinetic energy over the short distance of cellular dimensions. From an energetic viewpoint they would be a preferential energy source. Yet, they are not used as far as we know either. Again, the basic problem may be control, as well as insufficient frequency, and inconsistency. Radiation and particle emissions occur in a random fashion from a decaying atom. The frequency, direction, and precise level of energy are all unpredictable. It is difficult to envision how any organism could control the decay in a way to obtain energy on a consistent basis. On Earth, organisms have developed mechanisms to avoid, tolerate, and repair damage caused by ionizing radiation. A prime example is *Deinococcus radiodurans*, which possesses a high redundancy of repair genes to cope with ionizing radiation and organic pigments to cope with UV radiation.

4.3 The Question of Entropy, Uniformity, and Origin

A qualitative and quantitative assessment of the various energy sources indicates the theoretical plausibility that several of them potentially could power living organisms on other worlds. From a purely energetic view, the thermotrophic organism appears to be most favored. Such an organism may have in fact developed on Earth, and may be the progenitor of the photoautotroph as suggested by Muller (1995). His proposed thermosynthesis scheme is certainly simpler than photosynthesis, and heat is a ubiquitous energy source. However, we do not observe the presence of thermotrophic life on Earth, even though it appears to be favored on the basis of theoretical energetic considerations. It may be present and yet undiscovered on Earth, but terrestrial life definitely prefers chemical and light energy. What is the basic difference between thermal gradients on one hand, and chemical and light energy on the other hand that could account for this observation? Thermal energy has a high degree of entropy – it is highly disordered. Chemical and light energy are highly ordered forms of energy. According to the second law of thermodynamics, the degree of disorder in a system as a whole has to increase spontaneously with time. Any form of life is highly ordered and complex, and living processes increase the order of the system further. In order to maintain the highly ordered state of a living system, some free energy has to be expended to increase the degree of disorder, because the overall entropy of the system and its environment has to increase. In terrestrial organisms this increase in entropy is achieved by giving off heat and waste products, which are highly disordered. Thus, using a highly disordered energy source to begin with is very inefficient. Much less of the total amount of energy obtained from a more highly ordered source has to be converted into the disorder required to obey the second law of thermodynamics. Clearly, then, entropy is a factor that needs to be considered when assessing whether alternative energy sources can be used to power an organism.

Another factor that needs to be considered is uniformity of intensity. An alternative energy source has to provide its energy in quanta suitable and manageable for the organism to use. High-energy particles as emitted from decaying atoms are inconsistent and unpredictable, for instance. It may be that living systems can evolve a capacity to harvest energy more easily when that energy comes in the specific and consistent quanta appropriate for the control of metabolic reactions in a reliable way.

One other factor that needs to be considered is the question of pre-biotic evolution. How difficult was it for the earliest organisms to develop a mechanism for harvesting a specific energy source? The molecular machinery that has survived in chemoautotrophs and photoautotrophs today is very complicated, hence highly unlikely to have been the earliest mechanism for energy extraction. At the origin of life, the mechanisms for harvesting energy must have been simpler, and quite possibly were dependent on sources other than those that are used today. Chemoautotrophy and photoautotrophy are now the dominant basis for life on Earth, presumably because of the abundance and efficiency of those sources of energy. However, there may have existed early bioenergetic mechanisms that possibly were outcompeted over evolutionary time by more efficient mechanisms.

4.4 Survey of Energy Sources in our Solar System

Energy sources are ubiquitous in our solar system. Solar radiation providing light and thermal energy is one of the dominant energy sources for the interior terrestrial planets, and still significant for the Jovian and Saturnian systems. Heterogenous surface colorations such as on Venus and many of the icy outer satellites imply the presence of chemical energy. Geothermal energy is indicated by volcanism and a high-density interior that would imply radiogenic heating on planetary bodies such as Venus, Mars, and Io. Pressure is an energy source for planets with thick atmospheres such as the gas giants and Venus. Kinetic energy is a force on any planetary body with strong convection cells such as the gas giants, and possibly Europa. Tidal flexing is an energy source commonly found in the solar system, for example at Io, Europa, and Triton. Strong radiation and magnetic fields emanating from Jupiter and Saturn affect their satellites. Tectonic stress is definitely observable on Io, and osmotic gradients may be present on icy moons with a subsurface liquid ocean such as proposed for Europa, Ganymede, Titan, and Triton. Observations indicating the presence of various energy sources on the major planetary bodies of our solar system are listed in Table 4.1. This does not mean that the potentially available energy sources are actually used by putative organisms, but only that they are present based on our current knowledge. A discussion on the suitability of life on those planetary bodies based on energy sources plus other geoindicators is provided in Chap. 9.

4.5 Chapter Summary

We have used relatively straightforward evaluations and calculations to show that life does not need to be restricted to energy from chemical bonds and light as a basis for bioenergetics, as it apparently is on Earth. While light and chemistry do indeed represent cosmically abundant and efficient sources of energy, on many worlds other sources of energy may be more practical. Thermal, ionic, and osmotic gradients, and the kinetic energy of fluids in motion, appear to be the most promising alternatives to light and chemistry on worlds where the latter energy sources are available. But in specialized circumstances, other exotic forms of energy could be favored. Life evolving under those alternative conditions would be pressured by natural selection to make use of the forms of energy available. The numerous alternative options that are plausible within our own solar system are shown in Table 4.1. Even though terrestrial life may not use any energy source other than chemistry and light, the sensitivity of terrestrial organisms to many other forms of energy provides a hint of other bioenergetic possibilities elsewhere in the universe.

Table 4.1 Energy sources in the solar system that would be available in principle for other possible forms of life.

Body	Observations	Energy Source
Mercury	Thermal gradients pass through range for liquid water at slowly moving terminator. Possible water ice at poles. High density and electromagnetic field $\Rightarrow$ metallic core Heavy cratering	Solar radiation Geothermal
Venus	Extensive resurfacing $\Rightarrow$ geological activity and chemical recycling. Very dense atmosphere $\Rightarrow$ pressure gradients	Solar radiation Geothermal Chemical Pressure
Moon	Extensive cratering $\Rightarrow$ old surface Evidence of past volcanism High density $\Rightarrow$ metallic core. Strong sunlight. Minor amount of frozen water at poles	Geothermal Solar radiation
Mars	Daily to seasonal temperature fluctuations above melting point of water at some latitudes, volcanoes. High density and weak electromagnetic field $\Rightarrow$ metallic core. Surface erosion by flowing water. Likely aquifer beneath permafrost	Solar radiation Geothermal Chemical?
Jupiter Saturn Uranus Neptune	Gas giants with indistinct high-pressure atmosphere/liquid transitions, extensive turbulence. Strong magnetospheres and radiation output	Pressure Convection Magnetism Radiation
Io	Volcanism, extensive resurfacing, large size, and density > 3.5. Strong Jovian radiation Surface coloration $\Rightarrow$ complex chemistry. Weak intrinsic electromagnetic field	Geothermal Tidal flexing Chemical Stress (tectonic) Magnetism
Europa	Extensive resurfacing, density > 3 Surface coloration $\Rightarrow$ complex chemistry and chemical recycling Water ice surface, liquid subsurface water. Strong Jovian radiation. Projected high salt contents in subsurface liquids	Geothermal Tidal flexing Magnetism Chemical. Radiation Convection Osmotic gradients

Table 4.1 (continued)

Body	Observations	Energy Source
Ganymede	Extensive resurfacing ⇒ geological activity Rippled surface, low density, weak magnetic field Surface coloration ⇒ complex chemistry Strong Jovian radiation	Geothermal Tidal flexing Chemical Radiation Magnetism
Callisto	Low density ⇒ mostly water-ice Strong Jovian radiation Extensive cratering, lack of magnetic field ⇒ little internal energy	Magnetism Radiation
Enceladus	Very low density and high albedo ⇒ mostly water-ice Strong Saturnian radiation Heavily cratered but resurfaced in part, with evidence of ice geysers ⇒ internal energy	Geothermal Tidal flexing Radiation Magnetism Convection
Iapetus	Low density and moderate albedo ⇒ mostly water-ice Dark leading edge ⇒ possible hydrocarbon chemistry Heavily cratered: little internal energy	Chemical
Titan	Dense, colored atmosphere ⇒ complex chemistry Density $\simeq 1.8$ ⇒ organic liquids and water-ice, with solid core Atmospheric protection from radiation	Chemical Geothermal
Triton	Surface coloration ⇒ complex chemistry Unusual surface features ⇒ internal energy Density $\simeq 2$ ⇒ rocky core Elliptical, retrograde orbit ⇒ tidal flexing and seasonal temperature fluctuations; volcanism Possible subsurface ocean	Chemical Geothermal Tidal flexing Stress (tectonic)
Pluto	Density ~ 2.1 ⇒ rock/ice mixture	Tidal flexing
Charon	Mix of light/dark features ⇒ complex chemistry	Chemical

References

Aekesson S, Morin J, Muheim R, Ottoson U (2001) Avian orientation at steep angles of inclination: experiments with migratory white-crowned sparrows at the magnetic North Pole. *Proceedings of the Royal Society of London*, Series B: Biological Sciences 268: 1907–1913.

Amend JP, Shock EL (2001) Energetics of overall metabolic reactions of thermophilic and hyperthermophilic Archaea and bacteria. *FEMS Microbiology Reviews* 25: 175–243.

Balashova VV, Zavarzin GA (1980) Anaerobic reduction of ferric iron by hydrogen bacteria. *Microbiology* 48: 635–639.

Baumstark-Khan C, Facius R (2002) Life under conditions of ionizing radiation. In: Horneck G, Baumstark-Khan C (eds) *Astrobiology: the Quests for the Conditions of Life*, Springer Publ., Berlin, Germany, pp 261–284.

Beatty JK, Chaikin A (1990) *The new solar system.* 3rd edition, Sky Publishing Corporation, Cambridge, Massachusetts.

Blakemore RP, Frankel RB (1981) Magnetic navigation in bacteria. *Scientific American* 245: 42–49.

Borucki JG, Khare B, Cruikshank DP (2002) A new energy source for organic synthesis in Europa's surface ice. *JGR-Planets* 107 (E11): art. no 5114.

Bräucker R, Cogoli A, Hemmersbach R (2002) Graviperception and gravireponse at the cellular level. In: Horneck G, Baumstark-Khan C (eds) *Astrobiology: the Quests for the Conditions of Life*, Springer Publ., Berlin, Germany, pp 287–296.

Brock TD, Gustafson J (1976) Ferric iron reduction by sulfur and iron-oxidizing bacteria. *Appl. Environ. Microbiol.* 32: 567–571.

Chyba CF, Thomas PJ, Brookshaw L, Sagan C (1990) Cometary delivery of organic molecules to the early Earth. *Science* 249:366–373.

Doelle HW (1969) *Bacterial metabolism.* Academic Press, New York.

Farkas I (1935) *Orthohydrogen, parahydrogen, and heavy hydrogen.* Cambridge University Press, Cambridge, UK.

Feinberg G, Shapiro R (1980) *Life beyond Earth – the intelligent Earthling's guide to life in the universe.* William Morrow and Company, Inc., New York.

Fox SW, Dose K (1977) *Molecular evolution and the origin of life.* Marcel Dekker, New York.

Frankel RB, Blakemore RP, Wolfe RS (1979) Magnetite in freshwater magnetotactic bacteria. *Science* 203: 1355–1356.

Gonzalez-Partida E, Birkle P, Torres-Alvarado IS (2000) Evolution of the hydrothermal system at Los Azufres, Mexico, based on petrologic, fluid inclusion and isotopic data. *Journal of Volcanology and Geothermal Research* 104: 277–296.

Grinspoon DH (1997) *Venus revealed: a new look below the clouds of our mysterious twin planet* Perseus Publishing, Cambridge, Massachusetts.

Gusev VA (2001) On the source of energy for survival and multiplication of heterotrophs in the absence of organic substrate: formulation of the hypothesis. *Biophysics* 46: 826–832.

Hagemann M, Schoor A, Mikkat S, Effmert U, Zuther E, Martin K, Fulda S, Vinnemeier J, Kunert A, Milkowski C, Probst C, Erdmann N (1999) The biochemistry and genetics of the synthesis of osmoprotective compounds in cyanobacteria. In: Oren A (ed) *Microbiology and Biogeochemistry of Hypersaline Environments.* CRC press, New York, pp 177–186.

Haldane JBS (1954) The origin of life. *New Biology* 16: 12–27, Penguin Books, Harmondsworth.

Ioale P, Gagliardo A, Bingman VP (2001) Further experiments on the relationship between hippocampus and orientation following phase-shift in homing pigeons. *Behavioural Brain Research* 108: 157–167.

Irwin LN, Schulze-Makuch D (2003) Modeling putative multilevel ecosystems on Europa. *Astrobiology* 3: 813–822.

Kargel JS, Kaye JZ, Head JW, Marion GM, Sassen R, Crowley JK, Ballesteros OP, Grant SA, Hogenboom DL (2000) Europa's crust and ocean: origin, composition, and the prospects for life. *Icarus* 148: 226–265.

Khurana KK, Kivelson MG, Stevenson DJ, Schubert G, Russell CT, Walker RJ, Polanskey C (1998) Induced magnetic fields as evidence for subsurface oceans in Europa and Callistro. *Nature* 395: 777–780.

Kivelson MG, Khurana KK, Russell CT, Volwerk M, Walker RJ, Zimmer C (2000) Galileo magnetometer measurements: a stronger case for a subsurface ocean at Europa. *Science* 289: 1340–1343.

Lovley DR (1991) Dissimilatory Fe(III) and Mn(IV) reduction. *Microbiological Reviews*, June 1991: 259–287.

Lovley DR, Philipps EJP, Lonergan DJ (1989) Hydrogen and formate oxidation coupled to dissimilatory reduction of iron or manganese by *Alteromonas putrefaciens. Appl. Environ. Microbiol.* 55: 700–706.

Madigan MT, Martinko JM, Parker J (2000) *Brock biology of microorganisms*, 9th edition, Prentice Hall, New Jersey.

Miller SL, Orgel LE (1974) *The Origins of life on the Earth.* Prentice-Hall.

Morowitz HJ (1968) *Energy flow in biology.* Academic Press, New York.

Muller AWJ (2003) Potential extraterrestrial habitats for thermosynthesizers. *Astrobiology* 3: 555–564.

Muller AWJ (1995) Were the first organisms heat engines ? - a new model for biogenesis and the early evolution of biological energy conversion. *Prog. Biophys. Molec. Biol.* 63: 193–231.

Muller AWJ (1985) Thermosynthesis by biomembranes: energy gain from cyclic temperature changes. *J. Ther. Biol.* 115: 429–453.

Neidhardt FC, Ingraham JL, Schaechter M (1990) *Physiology of the bacterial cell. A molecular approach.* Sinauer Associates, Inc. Sunderland, Massachusetts.

Oparin AI (1938) *Origin of life.* Dover, reprinted 1953, New York.

Rettberg P, Rothschild LJ (2002) Ultraviolet radiation in planetary atmospheres and biological implications. In: Horneck G, Baumstark-Khan C. (eds) *Astrobiology: the quest for the conditions for life.* Springer Publ., Berlin, Germany, pp 233–243.

Sagan C, Salpeter EE (1976) Particles, environments, and possible ecologies in the Jovian atmosphere. *Astrophys. J. Suppl. Ser.* 32: 624.

Schulze-Makuch D (2002) At the cross-roads between microbiology and planetology: a proposed iron redox cycle that could sustain life in an ocean – and the ocean need not be on Earth. *American Society of Microbiology (ASM) News* 68: 364–365.

Schulze-Makuch D, Irwin LN (2001) Alternative energy sources could support life on Europa. *EOS Trans.* AGU 82: 150.

Schulze-Makuch D, Irwin LN (2002a) Energy cycling and hypothetical organisms in Europa's ocean. *Astrobiology* 2: 105–121.

Schulze-Makuch D, Irwin LN (2002b) Reassessing the possibility of life on Venus: proposal for an astrobiology mission. *Astrobiology 2*: 197–202.

Schulze-Makuch D, Goodell P, Kretzschmar T, Kennedy JF (2003) Microbial and chemical characterization of a groundwater flow system in an intermontane basin of southern New Mexico, USA. *Hydrogeology Journal* 11: 401–412.

Shihira-Ishikawa I, Nawata T (1992) The structure and physiological properties of the cytoplasm in intact Valonia cell. *Jpn. J. Phycol.* (Sorui) 40: 151–159.

Stolz JF, Oremland RS (1999) Bacterial respiration of arsenic and selenium. *FEMS Microbiology Reviews* 23: 615–627.

Thompson WR, Sagan C (1992) Organic chemistry of Titan-surface interactions. Eur. Space Agency Spec. Publ. SP-338: 167–176.

Van Dover CL, Cann JR, Cavanaugh C, Chamberlain S, Delaney JR, Janecky D, Imhoff J, Tyson JA (1994) Light at deep sea hydrothermal vents. *EOS Trans.* AGU 75: 44–45.

White SN, Chave AD, Reynolds GT, Van Dover CL (2002) Ambient light emission from hydrothermal vents on the Mid-Atlantic Ridge. *Geophysical Research Letters* 29: art. no. 1744.

Wilmer P, Stone G, Johnston I (2000) *Environmental physiology of animals.* Blackwell Science, Oxford.

Zimmer C, Khurana KK, Kivelson MG (2000) Subsurface oceans on Europa and Callisto: constraints from Galileo magnetometer observations. *Icarus* 147: 329–347.

5 Building Blocks of Life

Life is based on complex chemistry yet only a few of all the available elements participate in most life-supporting reactions on Earth: carbon, nitrogen, oxygen, hydrogen, phosphorous, and sulfur. Of these, the most characteristic element of biological systems is carbon. In this chapter we will discuss why carbon is so favored by life on Earth and whether other elements could replace carbon in its dominant role on other worlds.

5.1 The Uniqueness of Carbon

Carbon is the universal building block for life as we know it. Its ability to form complex, stable molecules with itself and other elements, particularly hydrogen, oxygen, and nitrogen is unique. Organic chemistry involves millions of compounds. The simplest are the alkanes, with the general formula C_nH_{2n+2} (for $n = 1$, the compound is CH_4, or methane). Alkanes are converted to other compounds by replacing a hydrogen with other functional groups. The most important substitutions for biochemistry are -OH (alcohol), -CHO (aldehyde), -COO-R- (ester, R = alkyl group), -COOH (carboxylic acid), $-PO_4$ (organic phosphate), and $-NH_2$ (amine). An organic acid with both amino (NH_2) and carboxyl (COOH) groups is an amino acid, and polycondensation of amino acids leads to proteins, whose virtually infinite variety of shapes provide a vast repertoire of macromolecular complexity. Carbon atoms can also be arranged in a ring, as in cycloalkanes and aromatic hydrocarbons, rather than a chain. The basic structure of the aromatic hydrocarbons is the benzene ring, a resonant ring held together by π-bonding. This structure forms the basis for cholesterol and steroids, which are vital biochemical compounds for many cellular structures (such as membranes) and functions (such as hormone signaling). Rings are also formed by -O- bridges between carbon atoms (Fig. 5.1). The stability of these structures accounts for the fact that sugars occur naturally in this form.

The great variety of structures formed from carbon, from chains and rings to three-dimensional macromolecules, are mostly stable within a broad temperature range. The versatility of carbon is further enhanced by its ability to form double and triple bonds, which alters the chemistry and geometry of the molecule as well as its temperature-dependent fluidity. This ability of carbon to build an almost unlimited range of molecules can be attributed to various factors: (1) the stability of carbon macromolecules due to a carbon-carbon bond energy that is higher than that of any other non-metal, and is comparable to the strengths of carbon-hydrogen and carbon-oxygen bonds; (2) carbon's mid-range value of electronegativity that promotes the formation of primarily covalent bonds; and (3) high activation energies for substitution and bond cleavage reactions due to the absence of lone pairs or empty valence orbitals, thus enhancing the stability of hydrocarbons and halocarbons to water and oxygen (Sharma et al. 2002).

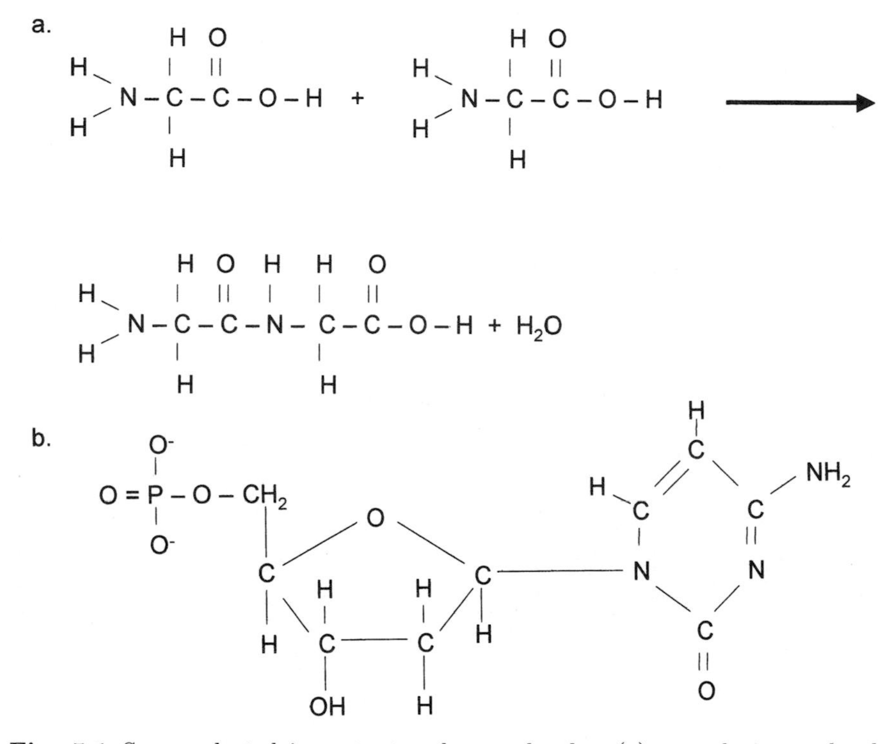

Fig. 5.1 Some selected important carbon molecules: (a) two glycine molecules (a simple amino acid) combine to form a peptide bond with the release of water as an important first step in the formation of proteins, (b) sugar (deoxyribose) – phosphate backbone and the base cytosine as one chain link of DNA (the next link of the chain would be the bonding of the upper oxygen of the phosphate group to the next sugar at the OH location).

Chirality is an important property of life-supporting carbon compounds. Only right-handed carbohydrates and left-handed amino acids are metabolically active in Earth's organisms. Enzymes, which regulate a variety of biological processes, have the ability to recognize the desired chirality of a substrate. This further enhances the versatility of carbon (though chirality is not limited to carbon compounds and occurs in most macromolecules).

The ability of carbon to form the backbone of long-chain polymers is central to its contribution to the chemical complexity of living systems. The hydrocarbons that make up the long chain and cyclic skeleton of lipids consist exclusively of covalent carbon-carbon bonds. The backbones of proteins are repeating units of two carbon atoms followed by a nitrogen atom. These chains are made possible because of similar bond strengths between a carbon-carbon bond and a carbon-nitrogen bond (Table 5.1). The backbone of nucleic acids is formed by repeating units of three carbon atoms, one oxygen atom, one phosphorus atom and another oxygen atom (Fig. 5.1). The capacity for each of these backbones to exist in a myriad of sizes and shapes, and to be modified by the placement of a great variety of reactive functional side groups, while maintaining the stability of the skeletal framework, provides living systems with an almost endless repertoire of stable but variable and interchangeable molecular forms.

Table 5.1 Typical bond energies for carbon and silicon with some other elements, as enthalpy of bond formation (kJ/mol).

Atoms	with Carbon	with Silicon
Hydrogen	435	393
Oxygen	$\sim$ 360	452
Nitrogen	$\sim$ 305	322
Chloride	351	381
Carbon	368	360
Silicon	360	340

Note: Bond strengths are highly variable depending on which compounds are involved and on whether single, double, or triple bonds formed. Data above indicate enthalpies of single bond formation. Data are from Greenwood and Earnshaw (1984).

Energetically favorable redox-reactions are the basis for metabolism. Carbon can be converted fairly easily between its fully reduced state (methane, CH_4) with a valence of -4 and its fully oxidized state (carbon dioxide, CO_2) with a valence of $+4$, which makes it a favorable element for use in metabolism. It is a considerable advantage that both products are gases under a fairly broad temperature range including Earth environmental conditions (Table 5.2). These compounds are the end products of metabolic activity – carbon dioxide for aerobic respiration and methane for methanogenesis – and because they are gases they can be disposed of readily as metabolic end products. Furthermore, they continue to be available for reactions. CO_2 dissolves easily in water and dissociates when exposed to short-wave UV radiation, and methane reacts with oxygenated compounds in the Earth's atmosphere.

Carbon appears to fit ideally with water as a solvent. Of the common elements, carbon has the strongest bond enthalpy to hydrogen and also forms a strong bond to oxygen (Table 5.1) – the constituents of water. Further, many biochemical reactions involve the dissociation or production of water, such as photosynthesis and the polycondensation of sugars, lipids, nucleic acids, and proteins. Carbon is also a relatively abundant element in the universe (Table 5.3), and thus no shortage of carbon as a universal building block of life should be expected in many types of planetary environments. Also, many complex carbon compounds have been found in the interstellar medium (Table 5.4) and meteorites (Table 5.5), further strengthening the dominance of carbon as a building block for life.

Table 5.2 Some physical properties of the fully reduced and fully oxidized forms of carbon and silicon.

Property	CH_4	CO_2	SiH_4	SiO_2
Molecular Weight	16.04	44.01	32.12	60.09
Melting Point (°C)	−182.5	−56.6	−185	1713^2
Boiling Point (°C)	−161.5	-75^1	−112	2950
Density (g/cm^3)	0.424 (at −164 °C)	1.03 (at −20 °C, 19.7 bar)	0.68 (at −186 °C)	2.65 (solid)

Notes: 1. Sublimation temperature at 1 bar, 2. SiO_2 as β-cristobalite. Data from Greenwood and Earnshaw (1984), Christen (1984), CRC (2001), and Air Liquide (2003).

Table 5.3 Some elemental abundances in the universe (as mass percent) and Earth. The elemental abundance of the Sun can be used as an approximate elemental abundance for the universe.

Atom	Sun	Earth	Earth's crust	Earth's Atmosph.	Ocean	Microbe
Hydrogen	91	< 0.1	0.14	< 0.001	11	63
Oxygen	0.08	50	46	21	86	26
Carbon	0.03	< 0.1	0.02	0.04	0.003	6.4
Nitrogen	0.010	< 0.1	0.002	78	5×10^{-5}	1.4
Silicon	0.003	14	28	< 0.001	2×10^{-4}	< 0.1
Sulfur	0.002	1.6	0.035	< 0.001	0.09	0.06
Phosphorus	3×10^{-5}	0.08	0.11	< 0.001	6×10^{-6}	0.12

Note: Data are from Goldsmith and Owen (2001) and CRC (2001).

Table 5.4 Some complex carbon compounds detected in the interstellar medium and meteorites.

Number of Atoms				
6	7	8	9	> 9
C_5H, HCH_2OH	CH_3C_2H	CH_3OCHO	$(CH_3)_2O$	$(CH_3)_2CO$
NH_2CHO,	CH_3CHO	CH_3C_3N	CH_3CH_2OH	HC_9N
CH_3CN	HC_5N, C_6H	C_7H, H_2C_6	CH_3CH_2CN	$HC_{11}N$
CH_3NC,	CH_3NH_2		HC_7N	C_6H_6, $C_{60}{}^+$
CH_3SH	CH_2CHCN		CH_3C_4H, C_8H	PAHs,
H_2C_4,	C_2H_4O		CH_3C_4N	glycine?
HCC_2HO,				
C_5H, C_5N,				
C_5O				

Note: Based on data from Chièze (1994), Goldsmith and Owen (2001), Ehrenfreund and Menten (2002), and http://www.spie.org/app/publications/magazines/oerarchive/july/jul97/extrater.html; PAHs = Polycyclic Aromatic Hydrocarbons.

Table 5.5 Some complex carbon compounds detected in the Murchison Meteorite.

> 1 000 ppm	100 – 1000 ppm	10 – 100 ppm	1 – 10 ppm
Sulfonic acids	Carboxylic acids	Amino acids	Phosphonic acids
	Polar hydrocarbons	Dicarboxylic acids	N-heterocycles
		Hydroxy acids	Purines
		Amides	Pyrimidines
		Alcohols	Amines
		Aldehydes	
		Ketones	
		Aliphatic hydrocarbons	
		Aromatic hydrocarbons	

Note: Data from Cronin (1998); ppm = parts per million.

5.2 Alternatives to Carbon as the Universal Building Block of Life

There has long been speculation about whether some other element could replace carbon as a universal building block for life (e.g. Scheiner 1891, Reynolds 1909, Spencer 1940). Boron, nitrogen, silicon, phosphorus, and sulfur are other common non-metallic elements that are known to form heat-resistant polymers. First, we will discuss silicon, which has many properties similar to carbon and probably is the most promising element to substitute for carbon. Then we will discuss briefly other elements that form polymers, which could alone or with other elements together form the backbone for polymeric molecules. It will be hard for any other element to match the complexity and versatility of carbon, but we have to keep in mind that (1) the research on polymeric chemistry is carried out overwhelmingly under Earth environmental conditions and many polymers that are stable below the freezing point or above the boiling point of water are unknown, and (2) life may not in principle require any particular element to be as dominant as carbon is in terrestrial biochemistry. Chains composed of just one element may not be necessary; chains of alternating atoms constructed from two or more different elements may work just as well (e.g. sequential units of B-N or Si-O). Proteins and nucleic acids already provide examples of structures that incorporate other elements into their backbones, as described in the previous section. Furthermore, in some other variety of carbon-based life, many of the carbon atoms could be replaced by silicon. The C-Si bond has about the same strength as the C-C bond (Table 5.1). Pure carbon-based and pure silicon-based forms of life may merely be the limiting cases, with a variety of transitional alternatives in between (Firsoff 1963). Some recent work on proteins and nucleic acids, where C atoms are replaced by Si atoms but functionality is attained, seem to support that assertion (Furusawa 1994, Chen et al. 2001).

5.3 The Possibility of Silicon-Based Life

5.3.1 Physical Properties of Silicon

Silicon is the most obvious potential substitute for carbon. It is also a p-block element of group IV (Group 14), just below carbon in the periodic table. With four electrons in its outer shell, it has somewhat similar physical properties to carbon (Table 5.6). Silicon and carbon are both small elements with small atomic weights and small atomic numbers, with carbon being the smaller of the two. Both elements have very high melting and boiling points, with carbon having the higher of the two (Table 5.6). Both elements are in the mid-range of electronegativities, but again carbon is higher. Both are solids at standard temperatures and pressures (STP) (298 K, 10^5 Pa) and both are semi-metallic. They both form sp^3 hybrid orbitals with tetrahedral structures in many of their compounds.

Silicon has a larger radius and therefore forms relatively weak bonds with the light elements that are abundant in the universe (Tables 5.1 and 5.3). The electrons in carbon are closer to the nucleus, and thus form stronger bonds that can retain light elements much better. This increases carbon's chance of forming complex compounds. The Si-Si bond strength is lower than the C-C bond strength, thus carbon is much more likely to bond with itself than silicon. The smaller Si-Si bond energy is also reflected in silicon's substantially lower energy of vaporization (Table 5.6). Silicon rarely forms any double or triple bonds, but double bonds and triple bonds are common and of great biological significance for carbon (Koerner and LeVay 2000). For example, amino acids, fatty acids, and nucleotides frequently have double bonds between carbon and oxygen. The fully oxidized form of silicon, silicon dioxide, forms four single bonds with four oxygen atoms resulting in a relatively inert mineral. Carbon dioxide, the fully oxidized form of carbon, forms double bonds with two oxygen atoms, resulting in CO_2, a reactive gas. Double and triple bonds are also used by organisms to store varying amounts of energy.

The homogeneous bond length of silicon is 235 pm compared to 144 pm for carbon, due mostly to silicon's larger electron cloud, which provides a greater magnitude of shielding and results in weaker Si-Si bonds compared to C-C bonds (Linn 2001). The larger size of silicon results in larger bond angles than for carbon (Zeigler and Fearon 1989), which has a great effect on which compounds can bond with silicon. For example, silicon cannot duplicate carbon's π-bonding, which is necessary to stabilize the delocalized electrons in resonant ring structures like benzene (Linn 2001). Thus, there is no set of aromatic compounds for silicon as there is for carbon.

Due to silicon's strong bond enthalpy with oxygen, it will be oxidized if oxygen is present. The high abundance of silicates on many of the rocky planets is caused by the bonding of silicon with oxygen when the planets formed. When silicon is fully reduced, it forms silane, a compound analogous to methane in the carbon system. Silanes burn spontaneously when in contact with oxygen to form a silicate and molecular hydrogen. Silane decomposes in the presence of water vapor to SiO_2, which explains why no SiH_4 has been detected in the hydrogen-rich atmosphere of Jupiter (though H_2O, CH_4, NH_3, PH_3 were detected) even though it has been detected in the interstellar medium (Goldsmith and Owen 2001). The much higher reactivity of silanes compared to the corresponding carbon compounds cannot be attributed only to lower bond strengths. Other factors include (1) the larger radius of silicon which is thought to facilitate attack by nucleophiles, (2) the great polarity of silicon bonds, and (3) the presence of low-lying d orbitals which permit the formation of 1:1 and 1:2 adducts[1], thereby lowering the activation energy of the reaction (Greenwood and Earnshaw 1984). The affinity of silicon to oxygen is so strong that if silicon is placed in water, it will form a silica shell, stripping the oxygen from the water (LeGrand 1998). Thus, water is not a compatible solvent for silicon-based compounds. Methane, ethane, or any compounds that contain methyl groups are more compatible solvents for a silicon-based system. Under surface conditions on Earth, the most stable silicon polymers are silicones, organic silicon polymers with a Si-O backbone. Silicon-silicon bonds are not stable under these conditions, but can be produced in the laboratory under conditions vastly differently from those of the

[1] unbounded association of 2 or 3 molecules in which a molecule of one compound, is either wholly or partly locked within the crystal lattice of one or two molecules of the other compound.

Earth's surface. A more thorough discussion of the polymeric chemistry of silicon and its possibilities for "exotic life" is provided in later sections of this chapter after we discuss the use of silicon by terrestrial organisms.

Table 5.6 Physical properties of carbon and silicon.

Physical Properties	Carbon	Silicon
Molecular Weight	12.011	28.086
Melting Point (in °C at 1 bar)	~ 3500	1414
Boiling Point (in °C at 1 bar)	~ 3900	3265
Density (g/cm^3 at 20 °C)	2.27[1]	2.34
Electronegativity	2.55	1.90
Single Bond Covalent Radius (pm)	77	118
Heat Capacity (J/g L at 25 °C)	0.709	0.705
Enthalpy of Fusion (kJ/mol)	0.00[2]	50.6
Enthalpy of Vaporization (kJ/mol)	394[3]	383

Note: 1. measured as alpha-graphite, 2. defined as standard state, 3. enthalpy of combustion. Data from Greenwood and Earnshaw (1984), Christen (1984), and CRC (2001).

5.3.2 Role of Silicon for Terrestrial Life

Silicon is not the basis for life on Earth, but it is still a very important component of living organisms. Without silicon, most of the plant and animal life on Earth would not exist as it does today. Plants use silicon as silica within the walls of the stems to provide rigidity so that the stem can remain erect, yet still remain flexible enough to not become brittle (Sangster and Parry 1981). Large amounts of silicon dioxide are used in the plant kingdom to form the rigid, defensive protrusions on plants. The silicon content of some plants such as the cucumber increases following fungal infection, and appears to exert a protective effect (Samuels et.al. 1991). Silicon, as silicic acid (0.1–0.6 mM), can be regarded as a plant nutrient (Epstein 1994, Birchall 1995). In the animal kingdom, silicic acid is a major constituent of the cells of the connective tissue where it is found in hair, nails, and the epidermis. Among vertebrates, silicon is essential for growth and bone development and for collagen and glycosaminoglycan synthesis (Carlisle 1981). When a bone is broken, high levels of silica are found around the break as it heals. The aorta, muscles, and tendons contain as much as 0.1% silicon, and the kidneys, heart, and liver contain about 0.05% silicon (Tacke and Wannagat 1979).

There is some evidence that silicon is an important nutrient for microbial life, and that it may even be used as a direct energy source for metabolism, or as a catalyst (Wainwright 1997). For example, Yoshino (1990) found that 100 μ g/mL of silicon has a remarkable stimulatory effect on the growth of

Staphylococcus aureus. Chakrabarty et al. (1988) and Das et al. (1992) showed that *Mycobacterium* and *Nocardia sp.* can grow in the absence of carbon, provided that silicon compounds are present. Similar observations were made by Tribe and Mabadje (1972) and Parkinson et al. (1989). They found that certain fungi grew in ultra-pure water only when silicon compounds were added. Wainwright et al. (1997) showed that silicic acid stimulates the growth of fungi, including *Penicillium* species, when growing in ultrapure water as well as nutrient-rich media. A possible explanation for these observations was provided early in the twentieth century. Reynolds (1909) suggested that silicon takes the place of carbon in some types of microbial metabolism, and Bastian (1914) suggested the existence of some form of silicon-based autotrophy, which, however, could not be confirmed. Lauwers and Heinen (1974) proposed that a silicon cycle operates in the environment involving microbial transformations between insoluble and soluble forms, which may in part be based on an earlier finding by Henderson and Duff (1965) that a wide range of bacteria and fungi can solubilize insoluble silicates by producing mineral and organic acids, and chelating agents.

One of the best examples of silicon use by a carbon-based form of life is the presence of silicates in the diatom (Fig. 5.2). Diatoms may account for as much as one-fourth of the world's entire net primary production, and serve as the basis for marine food webs. Although diatoms have a carbon-based energy metabolism, their shells are composed of silicon dioxide, and diatoms are absolutely dependent upon silicon for growth and development (Richter 1906, Lewin 1954). Diatoms require silicon in the form of $Si(OH)_4$ for their metabolism. Werner (1967) concluded that silicon affects (1) the citric acid cycle between acetyl-CoA and alpha-ketoglutarate, (2) the synthesis of special proteins, (3) the regulation of respiration, and (4) chlorophyll synthesis. Another example of the essential nature of silicon for diatoms is provided from an experiment conducted by Darley and Volcani (1969) in which DNA replication was stopped prior to mitosis by maintaining the newly divided cells of the diatom in a $Si(OH)_4$-free medium. The presence of silicon in the organelles of diatoms may indicate the possibility that it participates in the biochemistry of subcellular structures (e.g. Mehard et al. 1974, Azam et al. 1974).

Silaffins, a type of unique peptide, and long-chained polyamines are constituents of the biosilica of diatoms and can precipitate silica nanospheres (Kröger et al. 2002). The formation of the biosilica occurs within a membrane-bound compartment (silica deposition vehicle) that acts as a casting mold (Pickett-Heaps et al. 1990).

Thus, carbon compounds do interact with silicon compounds in diatoms to build a distinct organism. In diatoms the carbon compounds appear to act as a template for silicon structures. Cairns-Smith (1982, 1985) and Cairns-Smith and Hartman (1986) suggested that silicon compounds may have provided the first templates for carbon-based life, noting that the first carbon compounds

Fig. 5.2 Image of diatom *Asteromphalus hyalinus* provided by Tawnya Peterson, University of British Columbia (*top*), image of unknown diatom provided by Daniel Scheirer, Northeastern University (*bottom*).

may have received their initial handedness from clays or silicate minerals that also had handedness. For example, alpha-quartz, the thermodynamically most stable form of SiO_2 at Earth's surface conditions, forms tetrahedra that are interlinked to helical chains. Due to two slightly different Si-O distances of 159.7 pm and 161.7 pm and a Si-O-Si angle of 144°, the helices in any one crystal can be either right-handed or left-handed. A close association between minerals and the first forms of life is consistent with a variety of mineral characteristics (e.g. Table 2.1). Dessey (1998) suggested that silicon started the evolutionary process for carbon-based life on Earth. This is not impossible as conditions on Earth were quite different early in its history from today. Silicon could have played a vital role in the origin of life on Earth in its pre-biotic stage, but if so, its role was replaced by carbon relatively quickly as conditions became less harsh and more carbon friendly. If indeed silicon compounds were involved in molecular replication, this assembly alone must still have been far from the phenomenon that we call life, since this process meets only one of the characteristics of life as stated in Chap. 2.

5.3.3 Polymeric Chemistry of Silicon

Silicon hydrides or silanes have the generic chemical composition Si_nH_{2n+2} and occur both as branched and unbranched chains. They form direct analogs to hydrocarbons in the carbon-based world (Fig. 5.3). For example, a ring of six silicon atoms is called a cyclohexasilane, which has a direct analogy to a ring of six carbon atoms (Zeigler and Fearon 1989). However, because the silicon atom is larger than the carbon, it cannot duplicate carbon's π-bonding, which is essential for stabilizing the electron cloud in C_6H_6 and sustaining the benzene resonant ring structure. Thus, silanes do not form benzene analogs (siloxene has a ring of six silicon atoms but some of the Si-Si bonds are reinforced with oxygen bridges (Firsoff 1963), and thus are not an aromatic series analog). Because the bond energy between hydrogen and silicon is relatively weak compared to the bond energy to oxygen, silanes are much more reactive than their corresponding carbon compounds, and are readily oxidized into silicates in the presence of oxygen. The thermal stability of silanes decreases with increasing chain length. However, if hydrogen is replaced by organic groups, stable compounds are obtained. For example, polysilanes with molecular weights of above 10^6 have been synthesized (Sharma et al. 2002). Traces of alkali metals catalyze the hydrolysis of silanes, which is rapid and complete, to yield silicates and hydrogen molecules (Greenwood and Earnshaw 1984). Polysilanes are not stable at the temperature and pressure conditions of Earth's surface (with the exception of silane), but are fairly stable at low temperatures, especially at higher pressures.

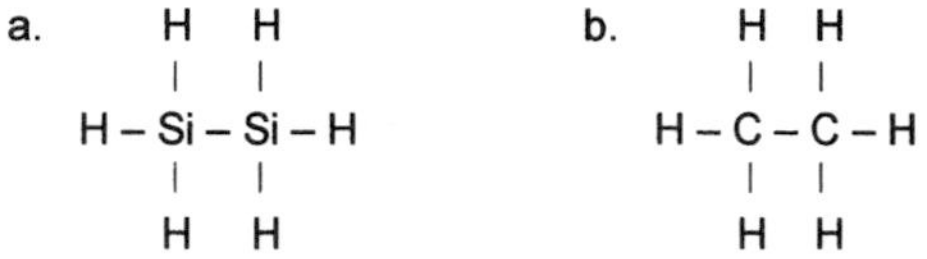

```
a.     H   H            b.     H  H
       |   |                   |  |
   H – Si – Si – H         H – C – C – H
       |   |                   |  |
       H   H                   H  H
```

Fig. 5.3 Structural analogy between (a) disilane and (b) ethane.

Replacing the hydrogen in silane with organic groups produces an analogue to hydrocarbons that eventually leads to silicon polymers, possibly resembling a reaction pathway to very complex silicon compounds. Organosilicon compounds (silicones) have considerable thermal stability (many up to about 400 °C) and chemical inertness reflecting the strength of both the Si-O bond and the Si-C bond (comparable in strength to the C-C bond) in the polymer backbone. Silicone polymers are relatively stable to oxidation, repel water, have good dielectric properties, and show a prolonged resistance to ultraviolet radiation (Greenwood and Earnshaw 1984). Hydroorganosilanes, organic compounds with silicon and hydrogen, have bonds that are thermodynamically more stable than other silanes, but become unstable at temperatures over 25 °C. The organosilicon reactions, for the most part, form directly without radical formation, while carbon tends to form radicals in the intermediate stages. Organosilicon compounds also more readily undergo nucleophilic substitutions when compared with their carbon equivalents. Yet, even with silicon's ability to produce complex molecules and carbon analogs, to more readily undergo reactions without the formation of radicals, and to take more direct routes in the formation of organic compounds, silicon accounts for less than one percent of all known organic compounds. Furthermore, the organic compounds of silicon that are known have all been produced in the laboratory but have not been found in nature (Pawlenko 1986). And while a very large number of compounds have been synthesized that involve multiply bonded silicon (e.g. West 1986, 1987, DeLeeuw et al. 1992, Stone and West 1994), stable compounds containing double or triple silicon bonds are difficult to form because of the larger atomic size and bond angles required (Zeigler and Fearon 1989). Silicon forms stronger bonds to nitrogen and the halogens than carbon does, but forms weaker bonds to hydrogen, phosphorous, and sulfur. The strong Si-O bond can be avoided and the carbon scenario reproduced if oxygen is replaced by sulfur. Then the resulting ratio of bonding energies of Si-Si to Si-S is comparable to the ratio of the C-C to C-O bonding energies (Firsoff 1963). Also, silicon polymers have been obtained with nitrogen instead of oxygen, where nitrogen acts as an electron donor. In hydrogen-poor environments, hydrogen is often replaced by a halogen such as chloride, and long linear chains of silicon and chloride are formed. Large labile molecules based on a Si-O-Si or Si-NH-Si backbone, with halogens as side-groups, could provide a basis for complex chemical systems.

Due to the strength of the Si-O bond, silicon forms a silicon tetrahedron in the presence of oxygen with the silicon in the center and a single bond to each of the four oxygens (Fig. 5.4). The tetrahedra can combine into chains, double-chains, rings, sheets and three-dimensional structures depending on the mineral formed. Neso-silicates consist of discrete SiO_4 tetrahedrons, soro-silicates of discrete Si_2O_7 units with one oxygen atom shared, cyclo-silicates of closed ring structures with two oxygen atoms shared, ino-silicates of continuous chains or ribbons with two oxygen atoms shared, phyllo-silicates of continuous sheets with three oxygen atoms shared, and tecto-silicates consist of continuous three-dimensional frameworks with all four oxygen atoms shared (Fig. 5.5). All silicates have in common high melting points (e.g. SiO_2 (β-cristobalite) at 1713 °C) and are therefore a major constituent of Earth's crust. Aluminum can substitute for silicon in the silicate network. At the surface conditions of Earth silicon dioxide is a tough, unreactive polymer (Koerner and LeVay 2000). However, at temperatures above 1000 °C complex silicate structures become more labile, partially melt and react with each other.

5.3.4 Environmental Conditions for the Possibility of Silicon-based Life

Based on the previous discussion, the prospect of silicon-based life under any conditions resembling those on Earth does not appear encouraging. Carbon can form a vast variety of complex compounds, not only from organic molecules on Earth, but also from material found in the heads of comets, inside meteorites, within the nebulae, and among the interstellar matter of the universe (Fegley 1987, Gladstone et al. 1993, Hanon et al. 1996, Llorca 1998, Varela and Metrich 2000). While polymeric carbon compounds seem to be ubiquitous, polymeric silicon compounds do not. Silicon, on the other hand, is found overwhelmingly as silicates making up the structure of the shell of the rocky planets, meteorites, and moons within our solar system. Recently, a very dense polymorph of silica has been discovered in the Martian meteorite Shergotty (Sharp et al. 1999). Although this polymorph could not support life, it still raises the question of what compounds silicon can and cannot form when subjected to conditions other than the constraints of Earth's surface. Thus, there is a need to discuss the chemical pathways of possible silicon-based life and the environmental conditions that could make it possible. Our approach is to distinguish between silanes, silicones, and silicates on the basis of the different environmental conditions at which these polymers are reactive. We start with silanes, because they present the closest analog to hydrocarbons, which are so important to life processes as we know them on Earth.

5.3.4.1 Life Based on Silane?

Silicon forms a series of hydrides, the silanes, consisting of Si-H and Si-Si single bonds (Fig. 5.4a). If silanes could be a basis for life as hydrocarbons are for carbon-based life, it would not be under the conditions found on present-day Earth, where silanes turn instantly into a silicate rock. A list of conditions can be derived that would have to be met as a minimum to make silane-based life a possibility (Table 5.7).

First, the atmosphere has to be reducing, with only minor amounts of oxygen available to avoid turning the silane polymers into silicate rocks. This kind of environment could exist if most of the original oxygen of a planetary body has been removed, for example by the precipitation of iron as iron oxides (as on early Earth). Oxygen could have also been lost to space due to fractionation during the early history of the formation of a planetary body and its atmosphere. Oxygen has a relatively high molecular weight of 32, but it dissociates to an atomic state in the outer atmosphere where the escape occurs. However, if the oxygen had escaped, then the much lighter hydrogen would also be expected to be severely depleted unless there was a specific mechanism to retain it. An oxygen-free atmosphere would allow the retention of free silicon instead of the formation of silicates.

Second, water is another compound that has to be scarce for silane-based life to form, because silicon is easily oxidized by water. Alkali-metals would likely be in solution catalyzing silane polymers into silicon dioxide and hydrogen gas. As for condition 1 (Table 5.7), the atmosphere would have had to undergo an extreme fractionation process to shed all lighter atmospheric gases with a molecular weight of up to 18 at least ($H_2O = 18\,g/mol$), or alternatively temperatures on the planetary body would have to be much below the freezing point of water to immobilize nearly all the water in the frozen state.

Third, temperatures far below the freezing point of water would be needed to make the silane-reactions that are so volatile at surface conditions on Earth controllable and accessible for life processes. A high-pressure environment would slow the silane-reactions as well, and thus have a complementary effect.

Fourth, and very importantly, a suitable solvent would be essential to foster the chemical reactions essential for life. Methane would be a good solvent for a silane-based system, and would have the additional advantage that it stays liquid at fairly low temperatures. However, methane has a relatively low molecular weight (16) and would be degassed if the atmospheric depletion of oxygen and possibly water is required (conditions 1 and 2; (Table 5.7)). Thus, in such an environment, heavier methyl-compounds such as methyl alcohol remain about the only alternative solvents.

Table 5.7 List of minimum conditions for silane-based life.

1. Little or no atmospheric or lithospheric oxygen
2. Little or no water in liquid form
3. Low temperatures (at least below 0 °C) and/or high pressures
4. Solvent suitable for silane-based complex chemistry
5. Restricted abundance of carbon

Fifth, the availability of carbon should be restricted, because carbon may be able to outcompete silicon for building complex macromolecules due to its greater chemical versatility. Other opinions on this issue have been offered, however. For example, Feinberg and Shapiro (1980) argued that the great versatility of carbon could also be considered a disadvantage because life-supporting molecules would have a very difficult time finding each other during the origin-of-life phase. Either way, some carbon may be an asset to silane-based life due to the possible formation of silicon-carbon bonds, but too much carbon appears to be a disadvantage.

The conditions imposed on the availability of silane polymers that could lead to a living system thus are very restrictive: a non-oxidizing, nonhydrous, extremely cold, high-pressure world where carbon is not abundant. While some of these restrictions are mutually compatible and commonly found in the universe, in their totality they are not found in our solar system and are not likely to be common elsewhere in the universe. The Saturnian moon Titan comes closest in our solar system to meeting the criteria. It meets all the criteria except for a low abundance of carbon, where the opposite is true. Still, there is some chance that under the extremely cold, reducing conditions on Titan, carbon would lose its competitive advantage over silicon.

The conclusion that silane-based life, if it exists, is not common in the universe is based on chemical reasoning, but is also supported by observational evidence. The apparent lack of silane polymers and the abundance of carbon polymers in meteorites would suggest that carbon-based life is much more probable than silicon-based life (Table 5.5). There are significant variations in chemical abundances in our galaxy and the universe (e.g. metallicity), but no evidence seems to suggest that our solar system is extraordinary. However, the possibility that life could exist in very exotic planetary environments based or partially based on silane-chemistry, as discussed for Titan above, should not be overlooked. The failure to detect large amounts of silanes in space may just mean that silanes are very rapidly oxidized to silicates under the conditions of interstellar space.

5.3.4.2 Life Based on Silicone?

A silicone is an organo-silicon polymer with a silicon-oxygen framework. Its simplest fundamental unit is $(R_2SiO)_n$ (Fig. 5.4b). Carbon atoms can be included in the chain. Silicones are thermally stable at much higher temperatures than silane, even at temperatures where hydrocarbons are not stable. Silicones are also resistant to oxidation and prolonged exposure to UV radiation, thus are durable polymers under Earth's surface conditions and at higher temperatures. However, they repel water and would need a solvent such as methane or a methyl-compound to exhibit polymer activity. There is no suitable solvent for silicone in appreciable amounts on Earth, thus carbon with its excellent solvent partner – water – undoubtedly had the edge for the evolution of life on Earth. This holds true even though silicon and oxygen are the most common compounds in the Earth's crust (Table 5.3). The fact that silicone-based life did not develop on Earth – or was outcompeted very early in the origin-of-life phase – dims the prospects for silicone-based life elsewhere. However, the main problem for silicone-based life on Earth may have been the abundance of water and the lack of a suitable solvent in appreciable quantities. An Earth-type world with methane as a major solvent is certainly imaginable, but whether this scenario would result in a silicone-based life form is very uncertain. Other planetary conditions that would make silicone-based life more likely are high pressures, higher temperature (50 – 400 °C), higher abundances of silicon than carbon, and a reducing atmosphere. A niche could exist at the temperature range of 200 – 400 °C, because carbon-based macromolecules generally disintegrate at about 200 °C and would not be competitive. But a solvent that would be liquid at that temperature is difficult to envision. Complex chemistry in a semi-molten state that would resemble life is conceivable, but the structural stability required for the low entropy requirements of a living system under such conditions seems improbable. Thus, the chemical properties of silicones render a silicone-based evolutionary pathway to biology unlikely.

5.3.4.3 Life Based on Silicate?

Silicates are salts containing anions of silicon and oxygen with the silicon-oxygen tetrahedron being the basic building block (Fig. 5.4c). At the surface conditions of Earth, silicates are inert, very slowly reactive polymers; but some silicates melt at temperatures beyond 500 °C, and most melt above 1000 °C where they become reactive. Feinberg and Shapiro (1980) suggested the existence of lavobes, organisms that could exist in lava flows, and magmobes, organisms that could exploit thermal gradients or chemical energy sources within the molten rock. These organisms could make use of the chemical complexity of silicate rocks in which aluminum could replace silicon in the tetrahedrons and cation exchange reactions could occur in interlayer sites between the tetrahedral and octahedral sheets. For this to occur, silicates would have to be in the form of sheets, such as the clay minerals smectite and montmorillonite. Information could be represented as irregularities in the crystal lattice of minerals (Cairns-Smith 1985). A particularly intriguing example is the clay mineral amesite, which has a helical structure. Information could be encoded by substitution of silicon with aluminum (Fig. 2.1).

It would be difficult to imagine how life could sustain itself at such high temperatures. Reactions at 1000 °C or above happen so fast that it would be difficult for an organism to control them. However, some silicates start to partially melt at much lower temperatures, such as zeolites. Recently, organic-inorganic hybrid zeolite materials have been synthesized where siloxane bonds (Si-O-Si) have been replaced with methylene frameworks ($Si-CH_2-Si$) (Yamamoto et al. 2003). Zeolite minerals could provide a suitable silicate membrane, because they have been shown to act as semi-permeable membranes comparable to cell membranes, preferentially filtering some molecules but not others (Falconer and Noble 2002). Thus, they possibly would be able to maintain disequilibrium conditions that are so crucial for life. They also may provide a durable enough membrane that would be needed as a boundary between the interior of the organism and the outside environment. Heron

(1989) suggested that zeolites may also be used by silicon-based life as enzyme mimics. Still, the problem of a suitable solvent at these temperatures, as discussed for silicone, remains.

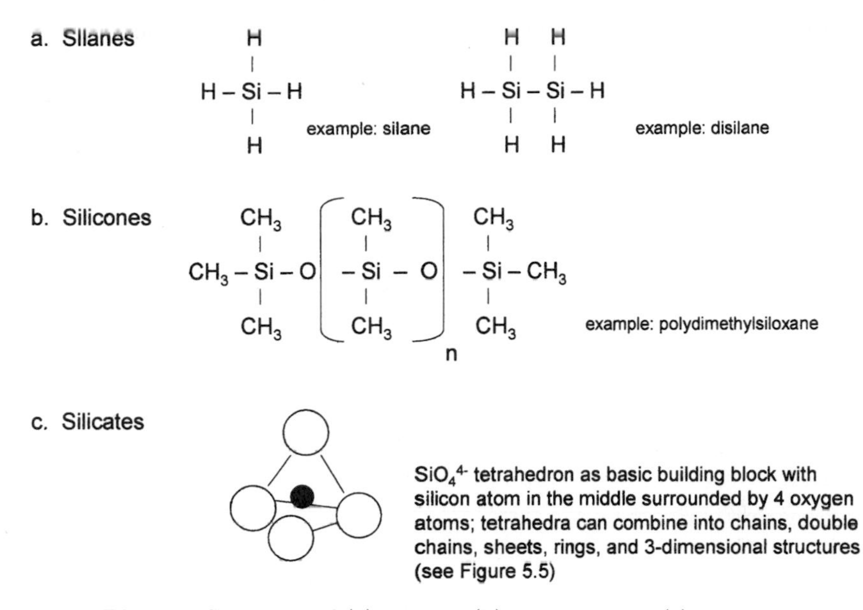

Fig. 5.4 Structure of (a) silanes, (b) silicones, and (c) silicates.

All inner planets of our solar system have magmatic activity in their interiors, as do some of the moons of the gas giant planets. For example, Io is the planetary body with the most volcanic activity in our solar system (Matson and Blaney 1999). All these bodies could present favorable environments for silicate-based organisms. However, given the environmental conditions of Earth's crust and its composition (46% oxygen, 28% silicon) with abundant silicate melts, Earth should be more favorable than most other planets and moons in our solar system for such a form of life. Yet no fossilized remnants or structures consistent with such an organism have been found, even though outcrops of igneous and volcanic rocks are abundant. Thus, the existence of such organisms seems very unlikely.

5.4 Other Alternatives as the Universal Building Block of Life

A few other potential substitutes for carbon deserve consideration as candidates for polymer-based complex chemistry. In order to be a viable alternative to carbon, the element should be a non-metal and be able to form at least

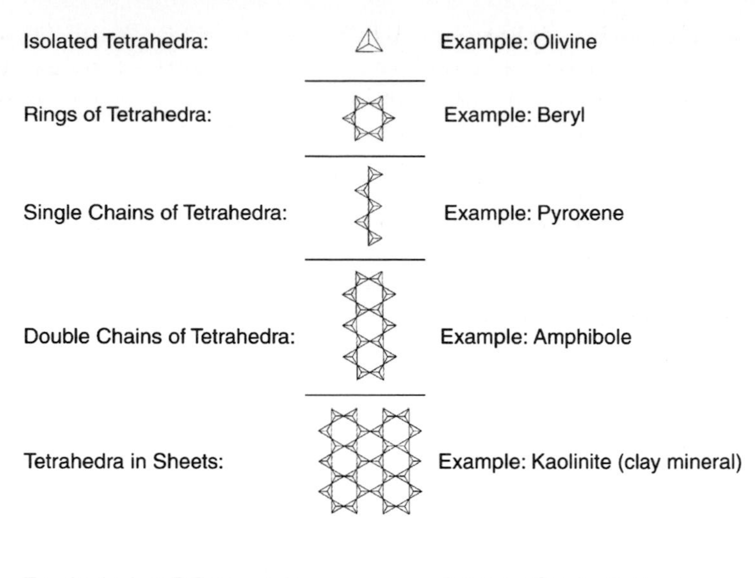

Fig. 5.5 Some silicate structures

the alkane-equivalent to the hydrocarbons. The possible choices are boron from group III of the periodic table, nitrogen and phosphorous from group V, sulfur from group VI, and possibly germanium from group IV (in addition to silicon).

Boron, like carbon and silicon, has a marked propensity to form covalent, molecular compounds, but it has one less valence electron than the number of valence orbitals. Thus, it has somewhat different chemical properties from carbon (Greenwood and Earnshaw 1984). Boron forms many different structural varieties of hydrides, but the boron atoms are linked indirectly through hydrogen bridges and are therefore not direct analogs to hydrocarbons. Boron has a high affinity for ammonia rather than water. It forms bonds with nitrogen that resemble the carbon-carbon bond. Two electrons of the nitrogen are donated in addition to the covalent electron sharing (Fig. 5.6). Boron-nitrogen compounds reproduce the physical and chemical properties of alkanes and aromatic hydrocarbons to a great extent, but with higher melting and boiling points. Borazole especially bears a strong resemblance to ordinary benzene in both physical and chemical properties (Firsoff 1963; Fig. 5.6). Borazole and its derivatives have a higher reactivity than the corresponding benzene group and thus would fit with the lower temperature range at which ammonia is a liquid solvent. Reactions at this lower temperature range would be at a more controllable pace. Also, boron has an affinity for nitrogen and ammonia as solvents that would fit into a low-temperature biological scheme.

Many B-N compounds, furthermore, exhibit high thermal stability. However, boron is an element of low abundance, with an average of about 10 ppm in Earth's continental crust (CRC 2001), thus a biological scheme based on boron without a strong fractionation mechanism appears unlikely.

Nitrogen can form long chains at low temperatures with a liquid solvent such as ammonia or HCN. However, the major drawback of nitrogen as a backbone for large molecular structures is that the energy of the triple bond in N_2 is much greater than that of the single bond, thus nitrogen-nitrogen bonds tend to revert back to elemental nitrogen. However, nitrogen can form longer molecular structures with boron as described above as well as with carbon, phosphorous, and sulfur. Nitrogen can also form hydrides for which hydrazine is an example. Phosphorus forms hydrides and has some merit as

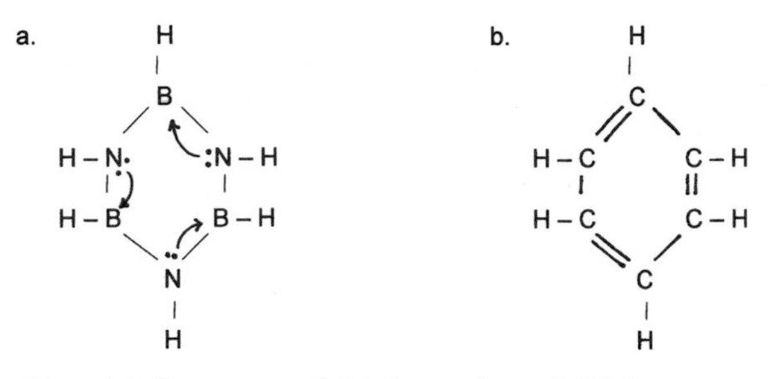

Fig. 5.6 Structure of (a) borazole and (b) benzene.

a potential building block with phosphine (PH_3) as compatible solvent. Sulfur forms hydrides as well, and could be of organic importance in a sulfur-rich environment with liquid solvents such as H_2S or H_2SO_4. Sulfur compounds with sulfur chains are known. However, given the poor variety of phosphorus and sulfur hydrides and the restricted environmental conditions, it is hard to imagine that either one of those elements would be a major building block of life. At best, they could form chains together with other elements such as carbon, silicon, or nitrogen. Sulfur, however, is unique in another way. It has more oxidation states than carbon, including fractional nominal states (e.g. +7, +5, +4, +3.33, +2.5, +2, −0.4, −0.5, −0.67, −1; Amend and Shock 2001), and thus can form a huge variety of different compounds. Germanium is not a suitable backbone element due to its large size, metallic character, and low abundance.

5.5 Chapter Summary

Carbon exhibits many characteristics that make it uniquely suited for life-supporting processes. Its usefulness for life derives primarily from (1) the versatility that enables it to form millions of complex polymers, including single-, double-, and triple-bonded compounds, chiral compounds, and resonant ring structures, (2) the ease with which it changes from one valence state to another, thereby suiting it well for energy-transferring redox reactions, and (3) its compatibility with water (and ammonia) as a liquid solvent. The only other element that approaches the versatility of carbon and is common enough to be a universal building block is silicon. Silicon can form long chains as silanes, silicones, and silicates. Some of these compounds could present a possible alternative to carbon for the construction of polymers under very restricted environmental conditions. These are (1) little or no oxygen, (2) little or no liquid water, (3) temperatures above 493 K (silicones, silicates) or below 273 K (silanes), (4) pressures greater than on the surface of Earth, (5) presence of a solvent such as methane or methanol, and (6) relative lack of available carbon. Other elements would likely not be suitable as backbones for the building blocks of a living system. However, it is not clear that polymeric skeletons have to be built from one element only. Polymers can also be assembled as chains of alternating elements such as Si-C, Si-O, and B-N. Alternation with carbon is used to some extent in terrestrial organisms (such as C-C-N in proteins and C-C-C-O-P-O in nucleic acids), and silated compounds play important structural roles in the cells of many organisms on Earth. But no comprehensive bioenergetic metabolism is known to arise from non-carbon complex chemistry, despite the high abundance of oxygen and silicon on Earth, and the relative concentration of silicon on other terrestrial planets. Thus, if elements other than carbon constitute the building blocks for any living system on other worlds, they almost surely exist under conditions far different from those on Earth, including temperatures and pressures where water could not be the solvent. Titan provides the best natural laboratory in our solar system for investigating this possibility.

References

Air Liquide (2003) http://www.airliquide.com/en/business/products/gases/gasdata, website, accessed 19 June 2003.

Amend JP, Shock EL (2001) Energetics of overall metabolic reactions of thermophilic and hyperthermophilic archaea and bacteria. *FEMS Microbiology Reviews* 25: 175–243.

Azam F, Hemmingsen BB, Volcani BE (1974) Role of silicon in diatom metabolism. VI silicic acid transport and metabolism in the heterotrophic diatom Nitzschia alba. *Arch. Microbiol.* 97: 103–114.

Bastian HC (1914) Experimental data in evidence of the present-day occurrence of spontaneous generation. *Nature* 92: 579–583.

Birchall JD (1995) The essentiality of silicon in biology. *Chem Soc. Rev.* 24: 351–357.

Cairns-Smith AG (1982) *Genetic takeover.* Cambridge University Press, London.

Cairns-Smith AG (1985) *Seven clues to the origin of life: A Scientific Detective Story.* Cambridge University Press, London.

Cairns-Smith AG, Hartman H (1986) *Clay minerals and the origin of life.* Cambridge University Press, London.

Carlisle EM (1981) Silicon in bone formation. In: Simpson TL, Volcani BE (eds) *Silicon and Siliceous Structures in Biological Systems*, Springer Verlag, New York, pp. 383–408.

Chakrabarty AN, Das S, Mukherjee K (1988) Silicon (Si) utilisation by chemoautotrophic nocardioform bacteria isolated from human and animal tissues infected with leprosy bacillus. *Indian J. Exp. Biol.* 26: 839–844.

Chen C-A, Sieburth S McN, Glekas A, Hewitt GW, Trainor GL, Erickson-Viitanen S, Garber SS, Cordova B, Jeffry S, Klabe RM (2001) Drug design with a new transition state analog of the hydrated carbonyl: silicon-based inhibitors of the HIV protease. *Chemistry and Biology* 8: 1161–1166.

Chièze J-P (1994) The interstellar medium. In: Audouze J, Israël G (eds) *The Cambridge Atlas of Astronomy.* Cambridge University Press, Cambridge, UK.

Christen HR (1984) *Chemie.* Verlag Diesterweg/Salle – Sauerländer, Frankfurt, Germany.

CRC (2001). *Handbook of chemistry and physics*, 82nd edition, Lide DR (ed), CRC Press, Boca Raton, Florida.

Cronin JR (1998) Clues from the origin of the solar system: meteorites. In: Brack A (ed) *The Molecular Origins of Life: Assembling Pieces of the Puzzle*, Cambridge Univ. Press, Cambridge, UK, pp 119–146.

Darley WM, Volcani BE (1969) Role of silicon in diatom metabolism. A silicon requirement for deoxyribonucleic acid synthesis in the diatom *Cylindrotheca fusiformis. Exp. Cell Res.* 58: 334–342.

Das S, Mandal S, Chakrabarty AN, Dastidar SG (1992) Metabolism of silicon as a probable pathogenicity factor for Mycobacterium and Nocardia. *Indian J. Med. Res.* 95: 59–65.

DeLeeuw BJ, Grev RS, Schaefer HF (1992) A comparison and contrast of selected and unsaturated hydrides of group 14 elements. *J. Chem. Ed.* 69: 441–444.

Dessey R (1998) http://www.sciam.com/askexpert/astronomy/astronomy28/, posted in Scientific American: Ask the Expert.

Ehrenfreund P, Menten KM (2002) From molecular clouds to the origin of life. In Horneck G, Baumstark-Khan C (eds) *Astrobiology – the Quest for the Conditions of Life.* Springer Publ., Berlin, pp 1–23.

Epstein E (1994) The anomaly of silicon in plant biology. *Proc. Natl. Acad. Sci.* (USA) 91: 11–17.

Falconer J, Noble RD (2002) Zeolite membrane research. http://www.colorado.edu/che/FalcGrp/research/zeolite.html.

Fegley B (1987) Carbon chemistry and organic compound synthesis in the solar nebula. *Meteoritics* 22: 378.

Feinberg G, Shapiro R (1980) *Life beyond Earth – the intelligent Earthling's guide to life in the universe.* William Morrow and Company, Inc., New York.

Firsoff VA (1963) *Life beyond the Earth.* Basic Books, Inc., New York.

Furusawa K (1994) Protection of nucleosides using bifunctional sully reagents. *Journal of the National Institute of Materials and Chemical Research* 2: 337.

Gladstone GR, Towe KM, Kasting JF (1993) Photochemistry in the primitive solar nebula; discussions and reply. *Science* 261: 1058–1060.

Goldsmith D, Owen T (2001) *The search for life in the universe.* 3rd edition, University Science Books, California.

Greenwood NN, Earnshaw A (1984) *Chemistry of the elements.* Pergamon Press, Oxford.

Hanon P, Chaussidon M, Robert F (1996) The redox state of the solar nebula; C and H concentrations in chondrules. *Meteoritics & Planetary Science* 31: Suppl. 57.

Henderson MEK, Duff RB (1965) The release of metallic and silicate ions from mineral rocks and soils by fungal activity. *J. Soil Sci.* 14: 236–246.

Heron N (1989) Toward Si-based life: zeolites as enzyme mimics. *Chemtech* Sept.-issue: 542–548.

Koerner D, LeVay S (2000) *Here be dragons – the scientific quest for extraterrestrial life.* Oxford University Press, New York.

Kröger N, Lorenz S, Brunner E, Sumper M (2002) Self-assembly of highly phosphorylated silaffins and their function in biosilica morphogenesis. *Science* 298: 584–586.

Lauwers AM, Heinen W (1974) Biodegradation and utilisation of silica and quartz. *Arch. Microbiol.* 95: 67–78.

LeGrand AP (1998) *The surface properties of silicas.* John Wiley & Sons, New York

Lewin JC (1954) Silicon metabolism in diatoms. I. Evidence for the role of reduced sulfur compounds in silicon utilization. *J. Gen. Physiol.* 37: 589–599.

Linn N (2001) Molecular visualization using methods of computational chemistry. Summer Ventures in Science and Mathematics, Univ. of N. Carolina at Charlotte, http://www.cmste.uncc.edu/papers/Can/%20Silicon/%20Based/2%0Life/%20Exist.doc

Llorca J (1998) Gas-grain chemistry of carbon in interplanetary dust particles; kinetics and mechanism of hydrocarbon formation. 29th Lunar and Planetary Science Conference 29: Abstract # 1119.

Matson DL, Blaney DL (1999) Io. In: Weissman M L-A, Johnson TV (eds) *Encyclopedia of the solar system.* Academic Press, New York, pp. 357–376.

Mehard CW, Sullivan CW, Azam F, Volcani BE (1974) Role of silicon in diatom metabolism. IV. subcellular localization of silicon and germanium. In: *Nitzschia alba* and *Cylindrotheca fusiformis*. *Physiol. Plant.* 30: 265–272.

Parkinson SM, Wainwright M, Killham K (1989) Observations on oligotrophic growth of fungi on silica gel. *Mycol. Res.* 93: 529–534.

Pawlenko S (1986) *Organosilicon chemistry.* De Gruyter, Berlin.

Pickett-Heaps J, Schmid AAM, Edgar LA (1990) The cell biology of diatom valve formation In: Round FE, Chapman DJ (eds) *Progress in Phycological Research* 7: Biopress, Bristol, UK, pp 1–169.

Richter O (1906) *Zur Physiologie der Diatomeen* (I. Mitteilung). Sitzber. Akad. Wiss. Wien, math.-naturw. Kl. 115: 27–119.

Reynolds JE (1909) Recent advances in our knowledge of silicon and its relation to organised structures. *Proc. R. Inst. UK* 19: 642–650.

Samuels AL, Glass ADM, Ehret DL, Menzies JG (1991) Distribution of silicon in cucumber leaves during infection by powdery mildew fungus (Sphaerotheca fulginea) *Can. J. Bot.* 69: 140–146.

Sangster AG, Parry DW (1981) Ultrastructure of silica deposits in higher plants. In: Simpson TL, Volcani BE (eds) *Silicon and Siliceous Structures in Biological Systems.* Springer Verlag, New York, pp 383–408.

Scheiner, J (1891) Astrophysicist at the University of Potsdam, considering life based on silicon in a scientific essay.

Sharma HK, Haiduc I, Pannell KH (2002) Transition metal complexes of germanium, tin and lead. In: Rappoport (ed) *The Chemistry of Organic Germanium, Tin and Lead Compounds* – Vol. 2, John Wiley & Sons, New York, Chap. 17, pp 1241–1331.

Sharp TG, El Goresy A, Wopenka B, Chen M (1999) A post-stishovite SiO_2 polymorph in the meteorite Shergotty: implications for impact events. *Science*, 284: 1511–1513.

Spencer JH (1940) *Life on other worlds.* Hodder and Stoughton, London, UK.

Stone FGA, West R (1994) *Advances in organometallic chemistry* – Vol. 36. Academic Press, New York.

Tacke R, Wannagat U (1979) *Syntheses and properties of bioactive organosilicon properties*, Springer-Verlag, Berlin.

Tribe HT, Mabadje SA (1972) Growth of moulds on media prepared without organic nutrients. *Trans Br. Mycol. Soc.* 58: 127–137.

Varela ME, Metrich N (2000) Carbon in olivines of chondritic meteorites. *Geochimica et Cosmochimica Acta* 64: 3433–3438.

Wainwright M (1997) The neglected microbiology of silicon – from the origin of life to an explanation for what Henry Charlton Bastian saw. *Society General Microbiology Quarterly* 24: 83–85.

Wainwright M, Al-Wajeeh K, Grayston SJ (1997) Effect of silicic acid and other silicon compounds on fungal growth in oligotrophic and nutrient-rich media. *Mycological Research* 101: 8.

Werner D (1967) Untersuchungen ueber die Rolle der Kieselsaeure in der Entwicklung hoeherer Pflanzen. I Analyse der Hemmung durch Germaniumsaeure. *Planta (Berlin)* 76: 25–36.

West R (1987) Chemistry of the silicone-silicone double bond. *Angew. Chem. Int. Ed.* 26: 1201–1211.

West R (1986) The polysilane high polymers. *J. Organometallic Chem.* 300: 327–346.

Yamamoto K, Sakata Y, Nohara Y, Takahashi Y, Tatsumi T (2003) Organic-inorganic hybrid zeolites containing organic frameworks. *Science* 300: 470–472.

Yoshino T (1990) Growth accelerating effect of silicon on *Pseudomonas aeruginosa* (in Japanese). *J. Saitama Med. Sch.* 17: 189–198.

Zeigler JM, Fearon FWG (1989) *Silicon-based polymer science: A Comprehensive Resource.* American Chemical Society, Washington, DC.

6 Life and the Need for a Solvent

Life as we know it consists of chemical interactions that take place in the liquid state, yet the requirement that life be liquid-based is not normally part of anyone's definition of a living system. Thus, we cannot state categorically that life in either a solid or gaseous state is impossible. There are, however, compelling theoretical advantages for the complex chemical interactions that compose the living state to occur in a liquid medium. These include (1) an environment that allows for the stability of some chemical bonds to maintain macromolecular structure, while (2) promoting the dissolution of other chemical bonds with sufficient ease to enable frequent chemical interchange and energy transformations from one molecular state to another; (3) the ability to dissolve many solutes while enabling some macromolecules to resist dissolution, thereby providing boundaries, surfaces, interfaces, and stereochemical stability; (4) a density sufficient to maintain critical concentrations of reactants and constrain their dispersal; (5) a medium that provides both an upper and lower limit to the temperatures and pressures at which biochemical reactions operate, thereby funneling the evolution of metabolic pathways into a narrower range optimized for multiple interactions; and (6) a buffer against environmental fluctuations.

For a substance to be an effective solvent for living processes, its physical properties in the liquid state must be matched to those of the environment in which it occurs. Those relevant properties include the requirement that it be liquid at the prevailing temperatures and pressures on the planetary body in question. These properties include the melting and boiling point of the solvent, but also its critical temperature and pressure. The critical temperature of a compound is that temperature beyond which the liquid phase cannot exist, no matter how much pressure is applied to it. The critical pressure of a substance is the pressure required to liquefy a gas at its critical temperature. A suitable solvent must also have sufficient physical buffering capacity, which can be specified by its enthalpy of fusion and vaporization (kJ/mol) describing the amount of energy needed to change 1 mol of the substance from solid to liquid at its melting point and from liquid to gas at its boiling point, respectively. A large temperature range for the liquid state is favorable. For those reactions that depend on the making and breaking of ionic and hydrogen bonds, and for maintaining appropriate macromolecular configurations, a measure of the solvent's polarity is important. The common measure is the dipole moment in debye (D, $1\,\mathrm{D} = 3.335 \times 10^{-30}\,\mathrm{Cm}$), which describes the polarity of a molecule and is dependent on charge and distance. The more polar the compound the higher is its dipole moment, but if the geometry of the molecule is symmetrical (e.g. CH_4, Fig. 6.1c), the charges are balanced and the dipole moment equals zero. The maintenance of appropriate diffusion rates depends on both density and viscosity. Viscosity, the quantity that describes a fluid's resistance to flow, is very much dependent on temperature and is measured in poise (P, dyne $\mathrm{s/cm^2}$) or pascal seconds (Pa s; $1\,\mathrm{Pa\,s} = 10\,\mathrm{poise}$). To the extent that electrical conductivity is relevant to a particular living process, the dielectric constant is also pertinent. The dielectric constant (dimensionless) is the ratio of the permittivity of a substance to the permittivity of free space and describes the extent to which a material concentrates electric flux (permittivity is the proportionality constant between electric displacement and electric field intensity). The physical properties for a variety of candidate solvents are provided in (Table 6.1).

First, we will discuss why water is such an excellent solvent for the environmental conditions of Earth. Then, we will discuss other possible solvents that could replace water under environmental conditions either similar to or vastly different from those existing on Earth. Finally, we will discuss how the nature of a solvent could determine the chemical characteristics of a living system, including the nature of its origin and evolution.

6.1 Water as the Universal Solvent for Life on Earth

Water is usually portrayed as the universal solvent for life as we know it, because of various properties that make it a very good solvent for the environmental conditions of Earth. Probably, the most important property of water is its polar structure (Fig. 6.1a). This polarity allows liquid water molecules to stick to each other via hydrogen bonding, providing it with polymer-like properties. The hydrogen bonding also raises the freezing and boiling point of water to much higher temperatures than would otherwise be expected from a molecule with a molecular weight as low as 18 g/mol. Water is a liquid in a temperature range of 0 °C to 100 °C (at 1 bar pressure). The triple point of water at which all three phases – solid, liquid, and gas – coexist is also located within this temperature range, very close to Earth's average temperature (Fig. 6.2). Thus, water in the liquid form allows for the wide variety of climatic conditions, differentiated habitats, and complex chemical and physical interactions found on Earth. And because life on Earth has evolved as a system that operates in liquid water, the temperature at which water is a liquid determines the range of temperatures at which living processes can proceed normally. The dynamic properties of life cease below the freezing

Table 6.1 Some physical properties of water and other polar compounds relevant to their solvent capabilities.

Property	H_2O	NH_3	HCN	HF	H_2S	CH_3OH	N_2H_4
Molecular Weight	18.015	17.031	27.02	20.01	34.08	32.04	32.05
Density (g/ml)	0.997	0.696	0.684	0.818	1.393	0.793	1.004
Melting Point (°C at 1 bar)	0.00	−77.73	−13.29	−83.35	−85.5	−94	1.6
Boiling Point (°C at 1 bar)	100.0	−33.33	26.0	20.0	−59.6	65	113.5
Range of Liquid (°C at 1 bar)	100	44.4	39.3	103.4	25.9	159	111
Critical Temp. (°C)	374	132	184	188	100	240	380
Critical Pressure (bar)	215	111	54	64.8	88	78	14.2
Enthalpy of Fusion (kJ/mol)	6.0	5.7	8.4	4.6	2.4	2.2	12.7
Enthalpy of Vaporization (kJ/mol)	40.7	23.3	25.2	30.3	18.7	40.5	40 900
Dielectric Constant	80.1	16.6	114.9	83.6	5.9	354(at 13 °C)	51.7
Viscosity (10^{-3} P)	9.6	2.7(at −34 °C)	2.0(at 20 °C)	~ 4.3	4.3(at −61 °C)	5.9	9.8
Dipole moment (D)	1.85	1.47	2.99	1.83	0.98	1.6	1.9

Notes: Data from CRC (2001), Dean (1992), Firsoff (1963), Merck Research Labs (1996), Moeller (1957), http://www.trimen.pl/witek/ciecze/old_liquids.html, and http://www.flexwareinc.com/gasprop.htm.

point of aqueous solutions, and are destroyed at temperatures above their boiling point.

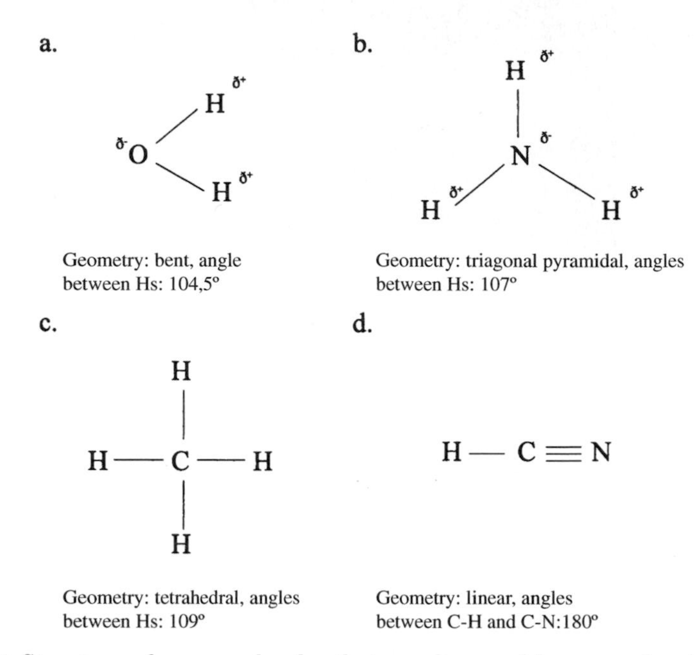

Fig. 6.1 Structure of some molecules that are discussed here as solvent: a. water, b. ammonia, c. methane, and d. hydrocyanic acid.

This empirical fact lends credence to the supposition that life is fundamentally a liquid-based process. Some tardigrades (hydrophilous micrometazoans) have been recorded to survive temperatures below $-250\,°C$ and as high as $151\,°C$ (Cavicchioli 2002), but this kind of survival occurs only in a dormant state.

The superiority of water as a good solvent for ionic and polar covalent bonds is a function of its high dipole moment (1.85 D), which quantifies the electric charges at its poles. Thus, water readily dissolves molecules such as salts that serve as cofactors for many metabolic reactions and mediate bioelectrical processes, as well as monomeric organic compounds with abundant polar groups such as sugars and amino acids that need to be capable of intracellular and transcellular mobility. On the other hand, water is not a good solvent for molecules with non-polar covalent bonds, such as those of hydrophobic organic molecules like lipids, which serve as the core of cellular membranes, and proteins embedded in the membrane core. Large biomolecules are thus able to maintain stable stereochemical configurations – a property essential for their biological activity – because of the stability of covalent carbonyl, peptide, glycosidic, phosphatidyl, and disulfide bonds

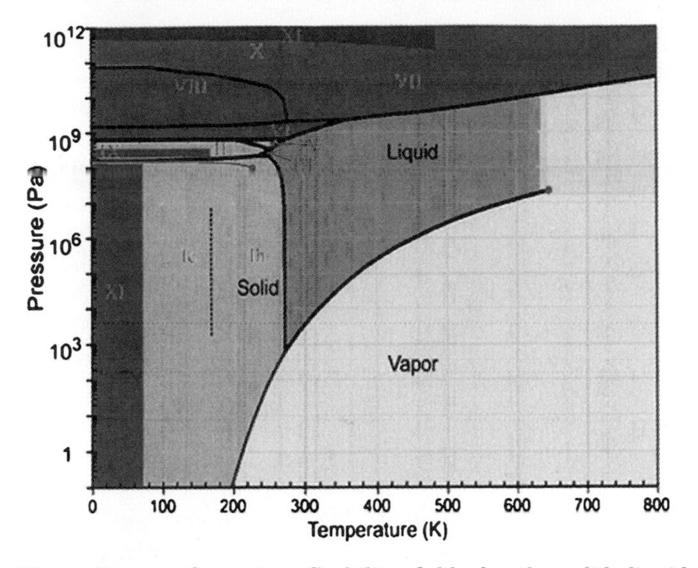

Fig. 6.2 Phase diagram for water. Stability fields for the solid, liquid, and gas phases as shown. All the different solid phases of ice involve hydrogen bonding of water molecules to four neighboring water molecules (from Chaplin 2003). Notice that the triple point is located very close to Earth's average temperature allowing all three phases to coexist under a wide variety of climatic conditions.

in an aqueous solvent. Another advantage of water is that it offers organic non-polar compounds a way to be concentrated.

There are challenges, however. First, many organic synthesis reactions involve dehydration, which is inhibited in the presence of water, hence requiring energetically expensive and elaborately catalyzed reactions. Second, water is very reactive and therefore damaging to many biomolecules, including high-energy phosphatidyl compounds that break down easily, and many cellular macromolecules, particularly the nucleic acids (Feinberg and Shapiro 1980). Specific repair mechanisms have evolved to counter the damaging reactivity of water with DNA, and metabolic evolution has evolved generally under the limitations imposed by water on biochemical processes.

One of the most biologically favorable properties of water is its role as a thermal moderator. The reason is that (1) water's high heat capacity allows it to be available over a wide range of temperatures (from 0 °C to 100 °C at 1 bar pressure), and (2) water is a good heat insulator. For example, the heat of vaporization, the amount of energy required to change from the liquid into the vapor phase, requires 40.7 kJ/mol (at a pressure of 1 bar), compared to 23.3 kJ/mol for ammonia and 18.7 kJ/mol for H_2S (Table 6.1). This high heat of vaporization stabilizes an organism's intercellular temperatures and promotes consistent chemical reaction rates. This same property also accounts for the cooling capacity of water evaporated from the surface of animals living in air, a mechanism important in the dissipation of excess heat.

The property of water to act as an insulator can be quantified by its dielectric constant, which is relatively high compared to most other potential solvents (Tables 6.1 and 6.2). Taken together, the wide temperature range at which water stays liquid and its insulating properties allow organisms immersed in liquid water to be very well protected from temperature fluctuations. This property is readily observable in the tendency of the oceans of the Earth to maintain moderate temperatures and counteract extreme temperature climatic fluctuations. Water also provides stability against environmental fluctuations in a chemical sense. Water dissociates into a proton (H^+), which is usually attached to a water molecule as a hydronium ion (H_3O^+), and an anion (OH^-), providing acid and base characteristics further increasing the solubility of compounds immersed in water. Drastic changes in pH, which could easily be detrimental to life, can be prevented by the presence of naturally and widely occurring buffers such as carbonate and phosphate. The presence of acid and base possibilities also leads to neutralization reactions, in which the cation of the base combines with the anion of the acid to form a salt, and the anion of the base combines with the cation of the acid to form a molecule of the solvent. Salts dissociate in their solvent, in this case water, into charged ions, increasing the reactivity of the solution as well as its conductivity. The usefulness of this electrolytic property is enhanced by the low viscosity of water. Ions can flow within and circulate efficiently through the solvent and participate in reactions within a reasonable time frame.

An immensely important property of water for all organisms is that it acts as a climatic stabilizer. Associated with that property is its unusual capacity to decrease in density as it freezes below its maximum density at 4 °C. This has obvious advantages for life in Earth's oceans and lakes, because the lighter water ice floats and acts as insulator for organisms beneath its surface and prevents the complete freezing of a body of water from the bottom to the top. This advantage is more important for multicellular forms of life than for microbial life, because microbes can more easily employ adaptive strategies against freezing such as surviving in a spore state. This peculiar property of water has a disadvantage as well, in that ice crystals pierce the cellular membranes due to the expansion of water as a solid. This is the basis for frost damage to plants, for example. Another peculiar property of water is its high surface tension, which reflects the tendency of water to form droplets due to greater cohesion among adjacent water molecules than between water and the molecules of air with which it is in contact. The surface tension of water is very high with a value of 71.99 mN/m at 25 °C (CRC 2001), compared to most other liquids. The surface tension of other potential solvents at their respective temperature of liquidity is not very well known, though. This property is not directly related to the solvent's ability to transport nutrients, but is likely to be relevant to the origin of life.

Organic compounds may have been concentrated in small water droplets that enhanced the probability of a reaction sequence leading to prebiotic molecules (Gusev 2002).

One important environmental advantage of water is that in an Earth-type environment it is self-shielding against ultraviolet radiation. Ultraviolet photons dissociate water molecules, releasing oxygen and hydrogen into the atmosphere. If that happens, some of the oxygen atoms will form ozone molecules, which are an excellent absorber for UV radiation, thus shielding the planetary surface from this detrimental form of high energy.

Another important property for any candidate to be qualified as a vital solvent in a scheme of organic chemistry leading to biology is that it has to be related to that scheme (Firsoff 1963). A solvent may be as good as water in an Earth-type environment, but if it does not form any ions that could advantageously enter into biochemical reactions, it is of little interest. With respect to water, hydration and dehydration reactions are widely in the biochemistry of Earth's organisms. Examples include the formation of peptide bonds between amino acids via dehydration synthesis reactions and the breakdown of complex sugars by hydration hydrolysis.

Finally, and perhaps most importantly, the solvent has to occur in large quantities on the planetary surface or wherever the life-sustaining reactions are to occur. With oxygen the most common element on Earth, and hydrogen the most common element in the universe, water would be expected to be common on Earth. Water is also known from spectral analysis to be a common molecule in the universe, and thus may be the obvious choice as a solvent elsewhere as well. However, many properties that make water such a good solvent are directly linked to the environmental conditions existing on Earth as discussed above. Thus, considering environmental conditions unlike those on Earth may require the participation of some other types of solvents. This notion will be explored in the coming sections.

6.2 Other Polar Solvents as Alternatives to Water

Several potential candidates could replace water as a polar solvent on other worlds. For example, Benner (2002) suggested sulfuric acid (H_2SO_4) as a possibility for Venus, and ammonia (NH_3) as a possibility for Jupiter. Most of the potential solvent candidates are liquids at lower temperatures than water. Hydrazine, N_2H_4, which is a liquid from 2 °C to 114 °C at 1 bar pressure, is one of the few exceptions. There is, however, overlap with the thermal range of liquidity for water in some cases (e.g. HCN, HF). Chemical reactions occurring within these solvents would proceed at a much slower pace than on Earth, typically by a factor of 2 for every 10 °C decrease in temperature (Jakosky 1998). However, the key to developing a suitable chemistry at a given temperature lies in selecting chemical reactions suited to that temperature. For example, those reactions involving unstable and highly reactive free radicals have very low activation energies (Feinberg and Shapiro 1980) and would be suited for low temperature organisms. Also, lower temperatures and an abundance of nitrogen could open up new possibilities, such as polymeric chains of nitrogen atoms. Bonds of nitrogen to nitrogen are weaker than carbon-to-carbon bonds, and are less abundant on Earth. Some of them are very reactive under Earth conditions (e.g. hydrazine). At lower temperatures, however, these compounds would be more stable and may be suitable for the construction of complex molecules. Many planetary bodies even within our solar system have much colder surface temperatures than Earth, and planets and moons at these temperature ranges may be much more common than Earth-type bodies in the universe.

In order to qualify as a suitable candidate for a polar solvent, the solvent has to be easily available and plentiful, as well as suitable for at least a hypothetical scheme that could lead to prebiotic chemisty. If not plentiful in the universe, there has to be some kind of fractionation mechanism that could conceivably enrich the particular solvent on a planetary body. In the following we discuss the polar solvent candidates ammonia, hydrocyanic acid, hydrofluoric acid, methanol, hydrazine, and possible sulfur-based solvents, which we rank according to the likelihood that they could replace water as the life-supporting solvent in certain types of environments.

6.2.1 Ammonia

The idea of life based on ammonia as a solvent has received a considerable amount of attention. For example, Haldane (1954) pointed out ammonia analogs to water and suggested the possibility of building proteins, nucleic acids, and polypeptides within a liquid ammonia solvent. Raulin et al. (1995) suggested that "ammono" analogs of terrestrial biomolecules in which oxygen atoms are replaced by NH groups might yield an equally viable pseudobiochemistry. Firsoff (1963) went into some detail showing similarities of synthesis reactions in water-based, ammonia-based, and water-ammonia mixtures. An especially interesting example provided by him is the synthesis of proteins from amino acids through a peptide bond. In a water system two glycine molecules combine with the release of water (6.1):

$$H_2N\text{-}CH_2\text{-}C(=O)\text{-}O\text{-}H + H_2N\text{-}CH_2\text{-}C(=O)\text{-}O\text{-}H \longrightarrow H_2N\text{-}CH_2\text{-}C(=O)\text{-}N(H)\text{-}CH_2\text{-}C(=O)\text{-}O\text{-}H + H\text{-}O\text{-}H \qquad (6.1)$$

If reaction (6.1) would take place in a water-ammonia mixture the COOH group would be replaced with a $CONH_2$ group and reaction (6.2) would be the result. In this reaction (6.2) the peptide bond is preserved and ammonia is released instead of water:

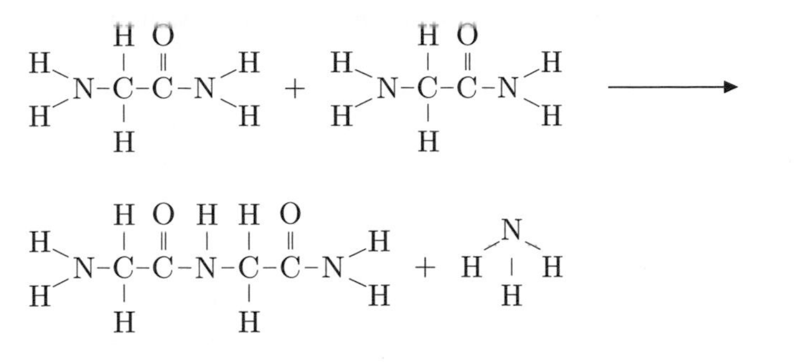

(6.2)

Firsoff interpreted that the preservation of the peptide bond may be a relic of an ammono-organic scheme on Earth in the early stages of evolution.

In a pure ammonia solvent without the presence of oxygen, the carboxyl group could be replaced with a -$CH(NH_2)NH_2$-group and the peptide bond with a -$CH(NH_2)$-NH- group. Similar analogs can be shown for the phosphate bonds in the nucleic acids (Firsoff 1963). A further parallel is that ammoniation reactions in ammonia as solvent are directly analogous to hydration reactions with water in which a salt and the solvent are produced. This leads to an important environmental consequence that minerals in a near-surface or subsurface "hydrosphere" would contain ammonia in their crystal structure just as Earth's rocks contain water.

Although there appear to be chemical pathways leading to prebiotic macromolecules, water is a more powerful solvent than ammonia based on its physical and chemical properties. Ammonia has a structure comparable to water (Fig. 6.1b) but a lower dipole moment and is thus less efficient at dissolving polarized compounds. Ammonia's heat of vaporization and heat of fusion are lower than those of water (although the heat of fusion is nearly equal, Table 6.1). Thus, ammonia is not as good at moderating temperature as water, but is still sufficient to have a stabilizing effect on climatic conditions. The surface tension of ammonia is only about one third that of water; thus ammonia is not likely to concentrate organic macromolecules in microscopic interfaces as well as water. Ammonia dissociates into nitrogen and hydrogen, and does not afford any protection from UV light when compared to the dissociation of water. Thus any origin of life in ammonia would have to occur in some kind of protected environment. Ammonia, however, makes up for this disadvantage by its ability to dissolve alkali metals without reaction, which is of biological relevance because alkali metals can act as catalysts. Also, salt solutions in liquid ammonia have usually a greater electrical conductivity than an aqueous solution of the same salt (Mee 1934). Ammonia is about four times less viscous than water, thus dissolved particles and ions have a less difficult time to find and react with each other. Ammonia self-dissociates as water does but to a much lower degree (1.9×10^{-33} at $-50\,°C$ compared to 10^{-14} at $+25\,°C$ for water). Ammonia dissociates to NH_4^+ (equivalent to H_3O^+ of water) and the anion NH_2^-. It can further dissociate to form two more base anions, NH^{2-} and N^{3-}, thus acid-base reactions do occur in an ammonia system but to a much lower degree.

Ammonia has a smaller range in which it stays liquid, thus life would have to adapt to a smaller temperature window to survive on such a world. Since solid ammonia is more dense than liquid ammonia, there is no preventive mechanism for preventing liquid ammonia from completely freezing during a cold spell. This is no problem for microbial life since it could have adapted to survive any cold spell or cyclic freezing in a spore state. However, it would make it unlikely that multicellular organisms could survive in such an environment. Further, chemical reactions would generally be expected to progress at a slower pace, which would likely have the result that evolution would progress at a slower pace as well. Microbial life in ammonia might not be as differentiated or as well adapted compared to water-based life on Earth. Further, liquid ammonia cannot co-exist with free oxygen, thus aerobic metabolism would be inconsistent with ammonia as a solvent.

Liquid ammonia could, on the other hand, present an opportunity for microbial life on the more numerous colder bodies in the solar system as well as for Jovian type planets where the boiling point of ammonia could be as high as ammonia's critical temperatures of $+\,132.4\,°C$ (given the extreme pressures). Ammonia and water are definitely related solvents, as indicated by the fact that life-sustaining organic macromolecules such as proteins, amino acids, and nucleic acids contain both OH and NH_2 functional groups in various combinations and proportions with which ammonia could easily interact. Interestingly, several enzymes of terrestrial organisms remain active down to temperatures of about $-100\,°C$ (Bragger et al. 2000). Fortes (2000) has suggested that in the likely subsurface oceans of the Galilean satellites, and possibly also at Titan and Triton, some ammonia dissolved in water may act as antifreeze, lowering the temperature at which water can stay liquid and thus possibly support life. Alternatively, any ammonia solvent in this type of environment should not be expected to be pure either. Water in the form of ice or icebergs would dissolve in liquid ammonia, thus the occurrence of water groups such as OH^- and O^{2-} should be expected within an ammonia solvent, if it indeed is capable of supporting life. Ammonia is certainly ubiquitous in the universe as it is found in the interstellar medium and in comets.

6.2.2 Hydrocyanic Acid

The structure of hydrocyanic acid, with a triple bond between the carbon and the nitrogen, is quite different from the structure of water, ammonia, and methane (Fig 6.1d). It is an excellent ionic solvent with a dipole moment of 2.8 D compared to 1.85 D for water and 1.47 D for ammonia. Also, as a thermal moderator it is about equal to water and superior to ammonia, with a heat of fusion of 8.41 kJ/mol and heat of vaporization of 25.2 kJ/mol, and a dielectric constant of 114.9. The range of temperature at which HCN remains a liquid is somewhat small, but it extends up to 26 °C allowing chemical reactions to proceed at a reasonable pace. Hydrocyanic acid dissociates into H^+ and CN^- with sulfuric and hydrochloric acids remaining acids in liquid HCN while cyanides are bases (Firsoff 1963). Cyan compounds are generally toxic to aqueous life in Earth's oceans and fresh water reservoirs. However, the toxicity only indicates the occurrence of reactions that are competitive with water-based biochemistry and the ease with which water groups can be replaced by CN. CN bonds are of fundamental importance in proteins and other organic substances, and the substitution of CN for OH would yield HCN-analog compounds. Some compounds valuable for life such as metals are insoluble or only soluble to a small extent in HCN, while other compounds such as certain salts (e.g. potassium thiocyanite, permanganate) are highly soluble and give highly conductive solutions (Moeller 1957).

Hydrocyanic acid is not as abundant as water in the universe but has been detected in comets and at planetary bodies such as Jupiter and Titan, and in the interstellar medium (Brown 1984, Lunine et al. 1999). Hydrocyanic acid not only offers protection from UV radiation, but can even combine with itself in alkaline solutions to form amino acids with the assistance of UV photons:

$$3HCN + 2H_2O + \text{UV light} \longrightarrow C_2H_5O_2N + CN_2H_2 \qquad (6.3)$$

In addition to glycine, this reaction produces cyanamide (CN_2H_2), which can link amino acids together as the first step in the formation of proteins. Another promising pre-biotic pathway was offered by Matthews and Moser (1966), who suggested the direct synthesis of protein ancestors (heteropolypeptides) from hydrogen cyanide and water without the intervening formation of amino acids. Also of interest is that the important purine adenine ($C_5H_5N_5$) is a pentamer of HCN. A biochemistry based on hydrocyanic acid would likely be quite different from one based on either water or ammonia, and that is perhaps the primary reason why HCN did not receive much consideration as a solvent for life. HCN cannot interact as well with Earth-type organic macromolecules as ammonia. Its relatively small thermal window of liquidity, and its limited ability to dissolve some biologically important compounds are disadvantages for HCN as a solvent for living systems. To the extent that Earth provides a natural laboratory for comparing HCN with H_2O, with temperatures covering the range at which both compounds are liquid, HCN clearly is not competitive with water as a solvent, either because of its lower abundance, or because the nature of biomolecules that would be stable and reactive in a hydrocyanic acid medium are not metabolically competitive with those that function in water. However, with its relative high molecular weight of 27 g/mol, it could be the solvent of choice on a planetary body where most of the lighter molecules may have steamed off at some stage of atmospheric evolution. In such an atmosphere, cyanogen (C_2N_2) could be expected as an atmospheric constituent, which might provide energizing reactions analogous to oxidation in Earth's atmosphere (Firsoff 1963).

6.2.3 Hydrofluoric Acid

The solvent properties for hydrofluoric acid and water are very similar. The dipole moment and the dielectric constant are basically the same, the heat of fusion and heat of vaporization of HF are only slightly lower (Table 6.1). Thus, hydrofluoric acid is an excellent temperature moderator and solvent. The temperature range at which HF remains liquid is a bit larger than for water and extends much lower (−83 °C to +20 °C). Two hydrofluoric acid molecules dissociate into HF_2^- and H^+. Due to their non-polar nature most hydrocarbon compounds are insoluble in hydrofluoric acid, but many of them are polymerized, decomposed, or lead to conducting solutions with complex cations, where the organic molecule bonds to the proton of HF (Firsoff 1963). Conceivably, F^- or HF_2^- could replace OH^-, and F_2^{2-} could replace O^{2-} in oxides. Fluorination could thus replace oxidation as the primary energy-yielding reaction, and would be more efficient due to the greater bonding energy of fluorine. As a result, free fluorine would be one of the atmospheric gases, and would likely have the same fate as oxygen had in the early Earth atmosphere (being such a very reactive element it would quickly be bound to surface rocks and removed from the atmosphere). The primary problem with hydrofluoric acid as a primary solvent is the low cosmic abundance of fluorine. Its abundance in the solar system was estimated to be about 1000 times less than that of carbon, nitrogen, or oxygen (Anders and Grevesse 1989). It is difficult to envision any fractionation mechanism that could enrich hydrofluoric acid to such an extent to become the primary solvent on a planetary body. One imaginable possibility is that all molecular oxygen could be used up in oxidation reactions on the planetary surface and that fluorine gas is released from F-rich magmas later in the history of the planetary body. However, due to its low cosmic abundance and the unlikelyhood of an efficient fractionation mechanism we rank HF as an alternative solvent lower than ammonia and hydrocyanic acid in spite of the advantageous properties of the hydrofluoric acid molecule.

6.2.4 Hydrogen Sulfide and Sulfur Dioxide

Hydrogen sulfide remains a liquid at colder temperatures than any other candidate solvent, but its temperature range as a liquid is only 26 °C. Hydrogen sulfide does not moderate temperatures very well, given its low heat of fusion, heat of vaporization, and dielectric constant (Table 6.1). It is not particularly efficient as an ionic solvent, given its low dipole moment, but it does dissolve many substances, including many organic compounds. Similarly to water, hydrogen sulfide dissociates into H^+ and SH^-. In a biochemical scheme with H_2S as solvent, the SH^- anion could simply replace the hydroxyl group in organic compounds. Hydrogen sulfide is a relatively common compound in the universe, usually associated with volcanic activity on planetary bodies. H_2S could be a conceivable solvent on Io, the volcanically most active planetary body in the solar system. A subsurface layer of hydrogen sulfide could turn liquid when overhead lava warms the subsurface layer up to its range of liquidity (Table 6.1), then "spores" could become activated, reproduce, and perpetuate an exotic subsurface microbial ecosystem. In this type of environment another sulfur solvent, sulfur dioxide, may compete with hydrogen sulfide. Sulfur dioxide is a solvent with a dipole moment of 1.6, remains a liquid at temperatures from $-75\,°C$ to $-10\,°C$, and could be retained preferentially on massive planetary bodies due to its high molecular weight of 64 g/mol. However, due to the double bond in sulfur dioxide the development of a biochemical scheme would be more complicated for SO_2 than for H_2S, because rearrangements would be needed. Also, SO_2 is not a proton-based solvent. Proton-based solvents have the advantage that organic macromolecules such as nucleic acids are constructed via hydrogen bonds and are able to exchange materials with the solvent or change their formation for biological purposes without having to overcome a high-energy barrier. Interestingly, sulfates including sulfuric acid (H_2SO_4) are insoluble in sulfur dioxide, and would thus be rock material in a pool of sulfur dioxide.

6.2.5 Methyl Alcohol (Methanol)

Hydrocarbon compounds are generally non-polar but can be made into polar molecules by replacing an H with an OH group in a hydrocarbon molecule. For example, methyl alcohol (CH_3OH) is an excellent polar solvent with a dipole moment of 1.68 D compared to 1.85 D of water and 1.47 D of ammonia. It is also a better temperature moderator than water based on its extremely high dielectric constant and heat of vaporization (Table 6.1) and remains a liquid in the wide temperature range from $-94\,°C$ to $+65\,°C$. It may also be a relatively common compound in the universe as it has been found in the interstellar medium and in comets (Goldsmith and Owen 2001). Liquid hydrocarbon compounds are known to exist on Titan in large quantities, but methyl alcohol does not appear to exist on Titan or elsewhere in the solar system in significant quantities.

6.2.6 Hydrazine

Hydrazine (N_2H_4) is a liquid at a temperature range from $2\,°C$ to $114\,°C$ and thus one of the very few solvent candidates that has a larger temperature range and a larger absolute temperature than water at which it stays a liquid. Hydrazine is an excellent polar solvent with a dipole moment of 1.9, comparable to water. Its dielectric constant (51.7) and viscosity (9.8×10^{-3} P) are also very similar to water. Based on its physical properties it would be an excellent solvent candidate as an alternative to water. Hydrazine, however, is a very reactive molecule and decomposes extremely fast in the presence of oxygen, making it an ideal rocket fuel. This property suits it poorly to serve as a solvent in the presence of oxygen. Based on its high reactivity, low temperatures in an anoxic environment perhaps would be conducive to controlled biochemical reactions at a reasonable speed. However, at temperatures below 2 °C hydrazine is a solid and therefore not a suitable solvent. Furthermore, hydrazine is not an abundant molecule, and thus does not appear to be a promising solvent candidate despite its favorable physical properties.

6.3 Non-Polar Solvents as Alternatives to Water

Membranes of terrestrial organisms, which are submerged in the polar solvent, water, are amphiphilic with their polar (hydrophilic) heads immersed in the solvent and their non-polar (hydrophobic) tails oriented toward each other, away from the solvent. They interact with the polar solvent to take up nutrients, respond to intracellular signals, and discard wastes. If a non-polar solvent could support life, the chemical orientation of membranes would have to be fundamentally different. In this case, the heads immersed in the solvent would have to be non-polar in order to interact with the hydrocarbon solvent.

A hydrocarbon solvent may actually improve chances for the origin of life, because many years of experience with organic synthesis reactions show that the presence of water greatly diminishes the chance of constructing nucleic acids. Thus, the assemblage of organic macromolecules that could give rise to life appears to be much more straightforward in a hydrocarbon environment. A hydrocarbon solvent would also provide protection against UV radiation, as hydrocarbon smog emanating from the liquid solvent would absorb some of the UV radiation and thus offer it a significant degree of protection. Hydrocarbon as the primary solvent on a planetary body is not outlandish. We can observe such a case at Saturn's moon Titan, in our own solar system. Spectroscopic results indicate the presence of methane rain on Titan as well as the presence of liquid lakes or even oceans of ethane on Titan's surface (Lunine et al. 1983, Lorenz 2000). A hydrocarbon liquid such as ethane, if it could support life as primary solvent, would likely produce organisms quite different from terrestrial organisms. If the hydrocarbon solvent is non-polar, such as methane or ethane, the organism's membranes would likely be hydrophobic on the outside, and hydrophilic or hydrophobic at their cores.

Methanogenesis could be an energy-yielding strategy in this type of environment. For example, photochemically produced acetylene, which is a solid under Titan's surface conditions (Lorenz et al. 2000), could be reduced with the help of hydrogen in the atmosphere to methane (6.4).

$$C_2H_2 + 3H_2 \longrightarrow 2CH_4 \tag{6.4}$$

In fact, methane is detected at a lighter isotopic fractionation than would be expected from Titan formation theory (Lunine et al. 1999), which could indicate activity of living organisms (Abbas and Schulze-Makuch 2002).

Radical reactions might also be used as metabolic pathway in a Titan environment, for example

$$\mathrm{CH_2\,radical + N_2\,radical \longrightarrow CN_2H_2} \tag{6.5}$$

$$\mathrm{2CH\,radicals + N_2\,radical \longrightarrow 2HCN} \tag{6.6}$$

Reactions involving radicals are very difficult to control and would cause internal damage to organisms on Earth, although some reactions involving intermediary radical products are catalyzed by terrestrial enzymes (Scott 2004). On Titan, however, at surface conditions of less than 100 K, these reactions should be better controllable, proceed at a reasonable pace, and may constitute a feasible energy-yielding reaction for a metabolic pathway. Reactions (6.5) and (6.6) have also the interesting side effect of producing the biologically important compounds cyanamide and hydrocyanic acid (Schulze-Makuch 2004, Schulze-Makuch and Abbas 2004).

Essential building blocks of life such as sugars, proteins and nucleic acids could exist in such organisms as well. However, given the vastly different environment from which that form of life would originate, it appears more likely that another solution to the make-up of life exists.

Table 6.2 provides some solvent-related physical properties of the non-polar hydrocarbons methane and ethane. They are common in the universe

Table 6.2 Comparison of physical properties for water and two non-polar compounds relevant to their solvent capabilities.

Property	H_2O	CH_4	C_2H_6
Molecular Weight	18.015	16.04	30.07
Density (g/ml)	0.997	0.426 at −164 °C	0.572 at −107 °C
Melting Point (°C at 1 bar)	0.00	−182	−172
Boiling Point (°C at 1 bar)	100.0	−161.5	−89
Range of Liquid (°C at 1 bar)	100	20.5	83
Critical Temp. (°C)	374	−82.6	+32.3
Critical Pressure (bar)	215	45.4	47.8
Enthalpy of Fusion (kJ/mol)	6.0	0.94	2.7
Enthalpy of Vaporization (kJ/mol)	40.7	8.2	14.7
Dielectric Constant	80.1	1.7	1.9
Viscosity (10^{-3} P)	9.6	0.009 at 20 °C	0.011 at 20 °C
Dipole moment (D)	1.85	0.0	0.0

Notes: Data from CRC (2001), Dean (1992), Goldsmith and Owen (2001), Firsoff (1963), and Moeller (1957).

and are primary liquid compounds on at least one planetary body of our solar system (Titan). These compounds are liquid at temperatures lower than water (as most of the polar substitutes for water are).

6.4 Quantitative Assessment of Solvent Candidates

In this section we will attempt to make a quantitative assessment regarding the suitability of a compound as an alternative solvent for life. There are some properties that are advantageous for life independently of the biochemistry of the living organism. Abundance is a definite advantage and may be the primary reason why life on Earth uses water as the universal solvent. Local abundance is most important for the solvent to be available for life processes, but overall cosmic abundance is relevant as well. If a compound is cosmically very rare a suitable fractionation mechanism has to be available to provide the compound in sufficient quantity. Aside from abundance, the most important property is being in a liquid state at the prevailing temperature of the local environment. The enthalpy of fusion and vaporization indicates how good a heat insulator the solvent is, and the dipole moment of the compound gives a general measure of its ability to dissolve other compounds (for life based on non-polar polymeric chemistry this criterion would be applied differently; for the present analysis that possibility will be disregarded).

A quantitative estimate of the feasibility that a given solvent could be effective in a particular planetary environment is shown in (Table 6.3). The metric is obtained by assigning +1 to four favorable characteristics – cosmic abundance, local abundance, enthalpy of vaporization, and dipole moment; by assigning −1 for unfavorable instances of the same characteristics; and by assigning 0 to cases that fall between clearly favorable and unfavorable characteristics. A fifth property – liquid at prevailing local temperature – is of such obvious importance that +2 is assigned for favorable cases, −2 for cases in which the solvent cannot be liquid at the prevailing temperature, and 0 for intermediate or unknown situations. The range of temperatures at which a solvent candidate is in the liquid state (at a pressure of 1 bar) is given in Fig. 6.3. Adding the assigned points for each of the five characteristics yields a single composite value for each solvent on each planetary body shown in Table 6.4.

This estimate is obviously crude and speculative to a large degree. A more precise and systematic analysis may emerge from further research, a consideration of other relevant variables, and a more highly differentiated system of weighting the different factors. Nonetheless, some generalizations can be gleaned from even this preliminary analysis. These include the inference that water is the best solvent on warmer bodies, but methanol may be superior on some colder worlds; that ammonia or ammonia-water mixtures (with ammonia being the "antifreeze" for water) may be an important solvent on the colder worlds as well (also supported by Leliwa-Kopystynski et al. 2002);

Table 6.3 Feasibility of a solvent for particular planetary environments.

Solvent	Earth					Icy Satellites					Gas Giants				
	C	L	T	E	D	C	L	T	E	D	C	L	T	E	D
H_2O	+	+	++	+	0	+	+	0	+	0	+	−	0	+	0
NH_3	+	−	−−	0	0	+	−	++	0	0	+	+	0	0	0
HCN	0	−	++	0	+	0	−	−−	0	+	0	−	0	0	+
HF	−	−	0	0	0	−	−	++	0	0	−	−	0	0	0
H_2S	−	0	−−	−	−	−	−	++	−	−	−	0	0	−	−
CH_3OH	0	−	++	0	+	0	−	++	0	+	0	−	0	0	+
CH_3OH_3	0	−	−−	−	−	0	−	++	−	−	0	−	0	−	−

Solvent	Io (near subsurface)					Titan surface					Titan subsurface				
	C	L	T	E	D	C	L	T	E	D	C	L	T	E	D
H_2O	+	−	0	+	0	+	0	−−	+	0	+	0	−−	+	0
NH_3	+	−	0	0	0	+	0	−−	0	0	+	+	++	0	0
HCN	0	−	−−	0	+	0	0	−−	0	+	0	0	−−	0	+
HF	−	−	0	0	0	−	−	−−	0	0	−	−	0	0	0
H_2S	−	+	++	−	−	−	0	−−	−	−	−	0	++	−	−
CH_3OH	0	−	0	0	+	0	0	0	0	+	0	0	++	0	+
CH_3OH_3	0	−	0	−	−	0	+	++	−	−	0	+	++	−	−

Cosmic Abundance (C): + = major component on planetary bodies or in interstellar space; − = found only in trace amounts
Local Abundance (L): + = major constituent in local environment; − = trace amounts or less in local environment
Thermal Range (T): ++ = liquid at prevailing local temperature; −− = liquid outside of prevailing local temperature
Enthalpy of Vaporization (V): + => 40 kJ/mol; − =< 20 kJ/mol
Dipole moment (D): + => 2; − < 1

Table 6.4 Summarized assessment of solvent candidates for selected planetary environments.

Solvent	Earth	Icy Satellite	Gas giants	Io (near subsurface)	Titan surface	Titan subsurface
H_2O	+4	+3	+1	+1	+1	0
NH_3	−1	+1	+2	0	0	+3
HCN	+1	−1	0	−1	0	0
HF	−2	−1	0	−2	−3	0
H_2S	−4	−3	−3	0	−4	−2
CH_3OH	+1	+1	0	0	+1	+2
CH_3OH_3	−4	−2	−3	−3	0	0

and that despite some favorable properties, hydrocyanic acid, hydrofluoric acid, and hydrogen sulfide do not appear particularly advantageous under any conditions.

With respect to specific planetary bodies, Table 6.4 suggests the following: (1) water is clearly the best solvent for Earth; (2) water is also the best solvent, perhaps in combination with ammonia and ethanol, beneath the surface of the icy satellites; (3) the gas giant planets do not favor any particular solvent except possibly water and ammonia within a particular thermal and pressure range; (4) the special circumstances of Io make it difficult for any solvent to function there, though some combination of water and H_2S might work beneath the surface; (5) Titan's surface is inhospitable for any other than organic solvents; while (6) the subsurface of Titan could function with a combination of organic solvents, ammonia, and water.

In conclusion,Table 6.4 suggests that other solvents may be more favorable under the many environmental conditions likely to be found on other worlds that differ from those on the Earth's surface. A good example is Titan, where water is in the solid state and ethane and methane are liquids, making them more suitable solvents. Over a large range of circumstances, however, it does appear that water is an adequate if not preferred solvent.

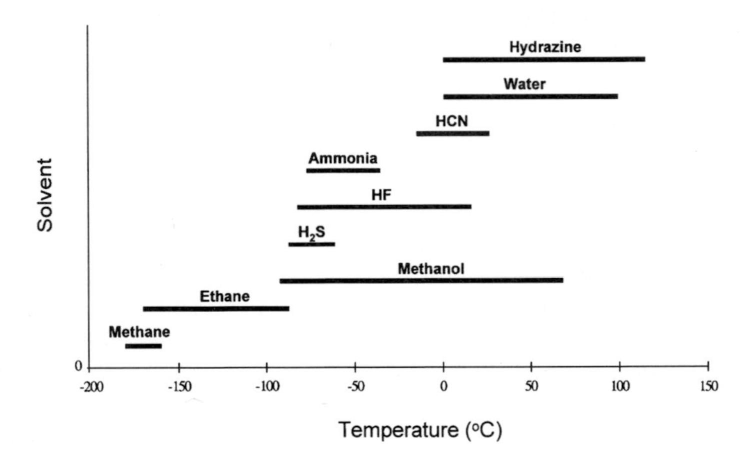

Fig. 6.3 Temperature ranges for solvent candidates to occur in the liquid state (at 1 bar).

6.5 Some Additional Thoughts

Except for the fundamental thermodynamics of its constituent molecules, there is probably no property of a living system that is more important for determining its characteristics than the nature of the solvent in which the system originated and evolved. Since the solvent determines the thermal limits within which chemical reactions can take place, the type of molecules that dissolve and do not dissolve, and the nature of the chemical interactions that can occur, both among the molecules and between them and the solvent, it follows that the total chemical makeup of the living system is a consequence of the solvent in which it occurs.

If our solar system is any indication, Earth-like planets are relatively rare, while planetary bodies that are either smaller and much colder, or larger and gaseous, are more frequently occurring. Both conditions render water useless as a solvent, but raise the prospects that an alternative solvent could be an effective medium for energy flow through chemically complex systems. Among the candidates are compounds that are not rare. Even those that are cosmically uncommon, like the sulfur-based solvents, may occur at critical densities under specialized circumstances (as on the volcanic planetoid, Io). In each case, however, the alternative solvent will almost surely foster a very different biochemistry from that which thrives in an aqueous environment. It is by no means apparent that exotic biochemical systems unfamiliar to us are less plausible under exotic conditions. As defined earlier, the essence of life has to do with complexity, energy flow, and information – none of which is limited in principle to the biochemical system with which we are familiar on Earth. In searching for life beyond Earth, we would therefore be well advised to expect the unusual.

The possibility does remain that the thermodynamic properties of any molecule that could undergo complex reactions and be assembled into stable macromolecular form is strictly constrained to conditions under which only water is an adequate solvent. Until research on complex chemical interactions in solvents other than water becomes more common, this possibility cannot be ruled out. If it should turn out that only water is an adequate solvent, the implication would be that life is likely to be restricted to water-abundant planetary bodies such as Earth, Mars, and Europa. Even if this were the case, the astronomical number of planetary bodies that surely exist in the universe, combined with the abundance of water throughout the cosmos, makes it possible that life even as we do know it may be widespread. Our hunch, however, is that the nature of life is broader and more pervasive than most of us imagine.

6.6 Chapter Summary

Water is an ideal solvent for the complex chemical systems that constitute life on Earth, because of its broad thermal range of liquidity corresponding to Earth's average temperature and atmospheric pressure, and because of other characteristics that make it both an excellent physical buffer and advantageous medium for complex chemical interactions among the biomolecules that have evolved under the physicochemical conditions prevailing on Earth.

On other worlds where temperatures and pressures are different, other solvents are more likely to be found in the liquid state. Many of them have properties that make them adequate, if not advantageous, as solvents for complex chemical systems. This includes, in particular, polar solvents like ammonia, hydrocyanic acid, and methanol. Under specialized conditions, others such as hydrogen sulfide and sulfur dioxide, could serve as solvents as well. Likewise, non-polar solvents like methane and ethane can be stable liquids under some conditions. While each of these could in principle host complex biochemical interactions, the nature of the molecules that would thrive in a given solvent would be dependent on the nature of that solvent. On worlds where the physicochemical conditions have mandated that a solvent other than water exists in liquid form, the conditions for the origin of biomolecules and the evolution of their variations and transformations would be different, and likely would result in the elaboration of a biochemical system different from that known to us from the example of life on Earth.

References

Abbas O, Schulze-Makuch D (2002) Acetyle-based pathways for prebiotic evolution on Titan. Proceedings of the 2nd European Workshop on Exo-Astrobiology (EANA/ESA), Graz, Austria, 16–19 Sept. 2002, ESA-SP. 518: 349–352.

Anders E, Grevesse N (1989) Abundances of the elements; meteoritic and solar. *Geochimica et Cosmochimica. Acta* 53: 197–214.

Benner S (2002) Weird life: chances vs. necessity (alternative biochemistries). Presentation given at "Weird Life" Planning Session for National Research Council's Committee on the Origins and Evolution of Life, National Academies of Sciences, USA, http://www7.nationalacademies.org/ssb/weirdlife.html.

Bragger JM, Dunn RV, Daniel RM (2000) Enzyme activity down to 100 °C. *Biochimica et Biophysica Acta* 1480: 278–282.

Brown RD (1984) Prebiotic matter in interstellar molecules. In: Papagiannis MD (ed) *The Search for Extraterrestrial Life: Recent Development.* D. Reidel Publishing Company, Dordrecht, pp 123–137.

Cavicchioli R (2002) Extremophiles and the search for extraterrestrial life. *Astrobiology* 2: 281–292.

Chaplin MF (2003) Professor at School of Applied Science, London South Bank University, http://www.sbu.ac.uk/water/phase.html, web site accessed 22 May 2003.

CRC (2001) *Handbook of chemistry and physics*, 82nd edition, Lide DR (ed) CRC Press, Boca Raton, Florida.

Dean J (1992) *Lange's handbook of chemistry.* 14th edition. McGraw Hill, New York.

Feinberg G, Shapiro R (1980) *Life beyond Earth – the intelligent Earthling's guide to life in the universe.* William Morrow and Company, Inc., New York.

Firsoff VA (1963) *Life beyond the Earth.* Basic Books, Inc., New York.

Fortes AD (2000) Exobiological implications of a possible ammonia-water ocean inside Titan. *Icarus* 146: 444–452.

Goldsmith D, Owen T (2001) *The search for life in the universe*, 3rd edition. University Science Books, Sausalito, California.

Gusev VA (2002) Chemical and prebiotic synthesis in the droplets of thunderstorm cloud. Proceedings of the 2nd European Workshop on Exo-Astrobiology, ESA-Sp. 518: 205–208.

Haldane JBS (1954) The origin of life. *New Biology* 16: 12–27,

Jakosky B (1998) *The search for life on other planets.* Cambridge University Press, Cambridge, UK.

Leliwa-Kopystynski J, Maruyama M, Nakajima T (2002) The water-ammonia phase diagram up to 300 MPa: application to icy satellites. *Icarus* 159: 518–528.

Lorenz RD (2000) Post-Cassini exploration of Titan: science rationale and mission concepts. *JBIS* 53: 218–234.

Lorenz RD, Lunine JI, McKay CP (2000) Geologic settings for aqueous organic synthesis on Titan revisited. *Enantiomer* 6: 83–96.

Lunine JI, Stevenson DJ, Yung YL (1983) Ethane ocean on Titan. *Science* 222: 1229–1230.

Lunine JI, Yung YL, Lorenz RD (1999) On the volatile inventory of Titan from isotopic substances in nitrogen and methane. *Planetary Space and Science* 47: 1291–1303.

Matthews CN, Moser RE (1966) Prebiological protein synthesis. *Proc. Natl. Acad. Sci. (USA)* 56: 1087.

Mee AJ (1934) *Physical chemistry.* William Heinemann, London.

Merck Research Labs (1996) *The Merck index*, 12th edition. Whitehousestation, New Jersey.

Moeller T (1957) *Inorganic chemistry*, 6th impression. Wiley, New York.

Raulin F, Bruston P, Paillous P, Sternberg R (1995) The low temperature organic chemistry of Titan's geofluid. *Adv. Space Res.* 15: 321–333.

Schulze-Makuch D (2004) Possible microbial habitats and metabolisms on Titan. Presentation given at "The Limits of Organic Life in Planetary Systems", National Academy of Sciences, USA, 10–12 May 2004, Washington D.C.

Schulze-Makuch D, Abbas O (2004) Titan: a prime example for alternative possibilities of life? In: Simakov M (ed) *Exobiology of Titan*, in press.

Scott J (2004) Co-Investigator of the NASA Astrobiology Institute, Carnegie Institution of Washington, Washington, USA, personal communication (provided at the Workshop on the Limits of Organic Life in Planetary Systems, National Academy of Sciences, 10–12 May 2004, Washington, D.C.)

7 Habitats of Life

There are four principal habitats in which life may exist – the surface of a planetary body, its subsurface, its atmosphere and space. From our own experience we know that life does exist on the surface of a planet, in its subsurface, and transiently at least in the atmosphere. Where it is present, it exists in a surprising diversity and in a variety of microhabitats, from deep caverns (Hose et al. 2000, Melim et al. 2001) to hydrothermal fluids and hot springs of various chemistries (Jannasch 1995, Rzonca and Schulze-Makuch 2002), to the frozen deserts of Antarctica (Friedmann 1982, Sun and Friedmann 1999). In this chapter we will elaborate on the principal habitats, the constraints they impose on life, and the possibilities they provide.

7.1 Life on the Surface

We live on the surface of our planet, which makes us biased toward it being the common case. However, there are various factors that make life on the surface of a planetary body challenging. Life on the surface is much more exposed to environmental extremes of temperature, wind, radiation, and humidity than, for example, life thriving in the subsurface protected by thick layers of soil and rock. A planet or moon with life on its surface requires an atmosphere to keep essential liquids on the surface from evaporating into the vacuum of space, to protect life on the surface from harmful cosmic and UV radiation (the degree of protection depending on the composition and thickness of the atmosphere), and to protect the surface to some degree from potentially devastating meteorite impacts. Smaller meteorites burn up in the atmosphere and the effect of larger ones is mitigated. However, meteorites still pose a grave threat to life on the surface of any planet. For example, the surface of our planet was most likely sterilized several times early in Earth's history (Sleep and Zahnle 1999). In that case life could have survived only deep in the crust and then resettled the surface again after the effects of the impact diminished with time. Life on the surface is also very susceptible to large-scale climatic fluctuations. Earth experienced several episodes of global freezing ("snowball earth") events (Hoffman et al. 1998, Kirschvink et al. 2000), when it substantially or perhaps completely froze over. Mars is currently a cold, arid planet with little or no liquid water on its surface, though it probably had oceans on its surface earlier in its history (Head et al. 1998, Dohm et al. 2000). Venus, probably wet and somewhat Earth-like early in its history, experienced a run-away greenhouse effect with current surface temperatures above 400 °C. If life as we know it ever existed on the surface of Venus, it does not anymore.

However, life on the surface does provide two critical advantages: (1) the use of visible light as an energy source and (2) space to expand. Life on Earth without photosynthesis would be much more limited and may have remained in the microbial evolutionary stage. The biomass contribution via photosynthesis is immense; the total carbon content of plants is estimated to be 560 Pg (1 Pg = 10^{15} g) for terrestrial plants and 1.8 Pg for marine plants (Schlesinger 1997, Table 7.1). Space to expand may not be very important to microbial life, because most microbial life easily fits into the pore spaces of rocks. However, for complex multi-cellular life the surface does provide a challenging but suitable environment for growth and development to macroscopic forms. Thus, it is not a surprise that we as macroscopic organisms populate a planet in a climatically fairly stable environment with enormous amounts of liquid water. However, life remains very vulnerable on a planetary surface. Threats include a large meteorite impact that could destabilize the climate or sterilize the surface of the planet, a cosmic disaster such as a nearby supernova-explosion that showers the surface with radiation, or exhaustion of fuel in the planet's central sun, leading to the engulfment of the planet as the sun expands to the red giant phase.

7.2 Life Beneath the Surface

Microbes, fungi, and small animals have lived in the upper layers of the soil since their first expansion from water onto the land. More recent evidence suggests that microbial life penetrates to great depths, beneath the surface of both the land and ocean bottom, deep into the crust (Pedersen and Ekendahl 1990, Johnson et al. 2003). Estimates indicate that the total amount of carbon in subsurface organisms may equal that of all terrestrial and marine plants (Table 7.1).

Table 7.1 Total carbon content in 10^{15} g of carbon.

Ecosystem	Plants	Soil and Aquatic Prokaryotes	Subsurface Prokaryotes
Terrestrial	560	26	22 – 215
Marine	1.8	2.2	303
Total	561.8	28.2	325 – 518

Note: Data from Whitman et al. (1998).

Microbial life appears to be abundant in various types of subsurface habitats such as the oceanic crust, and continental sedimentary and igneous rocks. While the overall number of organisms generally decreases with depth (Table 7.2), because of the huge amount of volume, the total subsurface biomass is enormous (Gold 1992).

Some distinct advantages over life at the surface can readily be appreciated. Temperatures and vapor pressures are stable, and protection from damaging radiation is afforded. The obvious disadvantages are the unavailability of sunlight as an energy source, and limitations on organismic size. The latter restriction results from the fact that pore spaces that serve as habitats are generally small in size, and become smaller with increasing depth. Thus, life at any substantial depth is probably restricted to microscopic dimensions, but this allows for a great range of living systems, as evidenced by the variety of microbial life within the crust of the Earth. Microbes indigenous to crustal rocks have been isolated from a depth of 2800 m in continental sedimentary rocks (Onstott et al. 1999) and 5300 m in igneous rocks (Pedersen 2000). Temperature increases with depth and imposes an absolute limit on the temperature and pressure conditions under which water can remain in the liquid state. The amount of dissolved solids in the ground water also tends to increase with increasing temperature adding osmotic stress to any organism. Thus, there is an absolute limit to the depth at which organisms can thrive. The absolute limit of this depth, however, is very variable due to the heterogeneous conditions in the crust and variable geothermal gradients.

Table 7.2 Total number of prokaryotes in unconsolidated subsurface sediments.

Depth Interval (m)	Cells/cm^3, $\times 10^6$	Deep Oceans (no. of cells, $\times 10^{28}$)	Continental Shelf and Slope (no. of cells, $\times 10^{28}$)	Coastal plains (no. of cells, $\times 10^{28}$)
0.1–10	220	66.0	14.5	4.4
10–100	45.0	121.5	26.6	8.1
100–200	6.2	18.6	4.1	1.2
200–300	19.0	57.0	12.5	3.8
300–400	4.0	12.0	2.6	0.8
400–600	7.8	NA	10.1	3.2
600–1200	0.95	NA	3.7	1.2
1200–2000	0.61	NA	3.2	1.0
2000–3000	0.44	NA	2.6	0.9
3000–4000	0.34	NA	NA	0.7

Note: Data from Whitman et al. (1998); NA = not available.

While sunlight is not available to provide energy in a subsurface environment, other sources of free energy are readily available. Chemical energy, both inorganic and organic, may be found in abundance, depending on the planetary body in question. Other energy sources such as those discussed in Chap. 4 may also be an option for certain subsurface environments. Availability of energy should not be a problem, if the planet or moon is large enough to have a metallic core and decaying radioactive elements as an energy source. Energy in many forms in principle can be transformed into biologically usable energy. If the availability of energy, then, is not an issue, and the living system is microscopic in size, the advantages of the subsurface habitat become overwhelmingly favorable for the persistence of life.

Although the subsurface clearly favors microbial life, there are a few niches and possibilities for macroscopic life. A bizarre example of macroscopic subsurface organism is a fungus of the *Armillaria* family, which is pathogenic to trees (Armillaria root disease). These fungi are incredibly large, with one *Armillaria ostoyae* organism of genetic uniformity recently detected at a size of 9.65 km^2 (Ferguson et al. 2003). A subsurface niche particularly favorable for macroscopic forms is the cave environment, to which various types of animals on Earth are ideally adapted. Caves do not occur only where karstic sedimentary rocks are present, but also commonly form in cooling lava flows. Thus, they can be expected to be common on other planetary bodies as well. Several locations have been suggested for Mars (Fig. 7.1). Due to the relatively low gravity on Mars, lava tube caves can be expected to be larger and more common than on Earth.

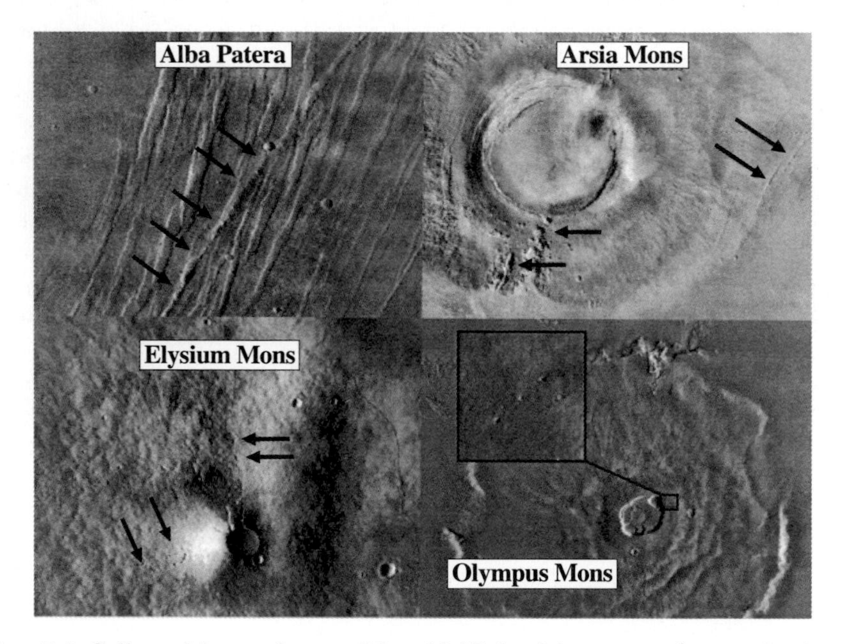

Fig. 7.1 Collapsed lava tubes on Mars highlighted by arrows (composite image provided by R.D. "Gus" Frederick, Silverton, Oregon, based on data from NASA).

On all the terrestrial planets and all the larger satellites, subsurface strata probably exist where thermal stability and some solvent in liquid form can exist. Thus, the presence of at least microbial life at multiple sites beneath the surface of planets and some of their satellites throughout the solar system is distinctly possible. The larger icy satellites that show evidence of tidal flexing or other energetic perturbations, such as Europa, Ganymede, Enceladus, Iapetus, Titania, and Triton, have at least the potential for liquid water beneath their icy crusts (Carr et al. 1998, Chyba 1997, Coustenis and Lorenz 1999, Khurana et al. 1998, McKinnon and Kirk 1999). Evidence for a substantial amount of ground water within the upper crust of Mars is now compelling (Boynton et al. 2002, Carr 1996, Malin and Edgett 2000, Schulze-Makuch et al. 2004a). Thus, aquatic life approximately as we know it on Earth is even possible in those situations. While the surface of Io is normally frozen, periodic lava flows heat it from above, and tidal flexing heats this rocky planetoid from below. Titan is colder still, but as the second largest satellite in our solar system, is obviously large enough for radiogenic heating to possibly liquefy mixtures of ammonia, water, and organic compounds which may be sequestered beneath its surface (Coustenis and Lorenz 1999). Both of these satellites represent more unusual but certainly possible subsurface habitats for life. Even Mercury and the Moon, both of which show evidence of some polar ice (Showstack 1998, Slade 1992), and Venus, where liquid silicates or water in a supercritical state could exist beneath the surface, cannot be completely ruled out as sites of possible subsurface life (Schulze-Makuch and Irwin 2002). "Run-away" planets that were ejected from their solar system and are now moving through empty space represent another theoretical possibility.

Stabilizing selection, which tends to narrow variation and optimize adaptive advantages (Chap. 3), is particularly effective in relatively constant habitats over long periods of time. Thus, life that has been optimized by stabilizing natural selection for a subsurface existence should be extremely durable and persistent. It also tends to be static, evolving little from the form and function that characterized it upon its introduction to the stabilized habitat. Subsurface environments may thus be repositories for early forms of life that have changed little since conditions made life on the surface untenable. On Earth, the microbes that are found at the greatest depths beneath the surface tend to be members of the evolutionarily ancient Archaea, or Eubacteria with ancestral forms of chemoautotrophic metabolism, and may include some types of nanobacteria. It follows that life on other worlds is most likely to be found beneath the surface of those worlds, where it is probably microscopic and relatively unchanged from an ancestral state.

7.3 Life in the Atmosphere

The possibility that the gaseous envelope of planets and those satellites that hold an atmosphere could serve as a suitable environment for life is generally viewed with skepticism. This probably derives from our familiarity with the nature of the atmosphere and of life on Earth. We are aware of the diversity of terrestrial life, both in the subsurface and on the surface of Earth, but no organism that spends its entire life cycle in the atmosphere has been documented. The lack of green clouds is visual evidence for the absence of concentrated airborne photosynthesis. And even the smallest organism has a higher density than air. Nevertheless, it has been recognized for some time that bacteria exist in cloud aerosols on Earth (e.g. Gislén 1948) and that rain and fog water rich in nutrients may provide a good substratum for microorganisms (Herlihy et al. 1987, Fuzzi 2002). While the dispersal of microorganisms by wind is well accepted (Schulze-Makuch 2003) and may even occur within the stratosphere (Imshenetsky et al. 1978), the claim that microbes independently grow and reproduce in Earth's atmosphere is controversial. Dimmick et al. (1979) reported the division of bacteria on airborn particles and more recently Sattler et al. (2001) analyzed condensing clouds at the Sonnblick Observatory in Austria at an altitude of 3106 m, and suggested growth and reproduction of microbes in super-cooled cloud droplets.

However, in general the atmosphere of Earth is a poor analogy for atmospheric habitats where life would be more likely; namely that of planetary bodies or satellites where gases are denser, and liquids are found in larger aggregates with longer survival times. Also, any particles in the Earth's atmosphere have typically a short residence time in the range of several days only. Most atmospheres of other planetary bodies we know are dynamically much more stable in the vertical direction and particles do not precipitate out as frequently (e.g. Venus), thus particle residence times are much longer. On other planets various chemical compounds might serve as nutrient sources, such as H_2S in the case of Venus (Schulze-Makuch and Irwin 2002) or complex carbon compounds in the case of gas giants like Jupiter (Boston and Stoker 1983, Stoker et al. 1990) or carbon-rich moons like Titan. The composition of some planetary atmospheres is provided in Table 7.3. Note the extremely significant recent detection of methane in the martian atmosphere.

If, instead of the unstable and thin atmosphere of Earth, the denser atmospheres of Venus, Titan, and the gas giant planets are taken as a prototypical atmospheric habitat where life could exist, some positive advantages can be noted. For instance, many of the denser atmospheres are more stable and more richly endowed with organic molecules. Sunlight, especially in the ultraviolet frequencies, breaks apart simple organic molecules in the planetary atmosphere, producing ions, free radicals, and other highly reactive molecules that combine to form complex, energy-rich compounds. These heavier molecules sink until they reach a level where they are destroyed by temperature and pressure, as probably occurs on the gas giants, or they ac-

Table 7.3 Composition of some planetary atmospheres (modified from Lewis 1995).

Planetary Body	Major Compounds	Minor Compounds	Trace Compounds
Venus	CO_2 (96.5%), N_2 (3.5%)	SO_2, Ar, CO, H_2O, He, Ne, H_2S	HCl, Kr, HF, COS
Earth	N_2 (78.1%), O_2 (20.9%), Ar (0.9%)	H_2O, CO_2, Ne, He, CH4, Kr	H_2, N_2O, CO, Xe, O_3, NH_3, SO_2, H_2S, CH_2O, NO_2, NO, HCl,
Mars	CO_2 (95.3%), N_2 (2.7%), Ar (1.6%)	O_2, CO_2, H_2O, Ne,	Kr, Xe, O_3, CH_4^*
Jupiter	H_2 (82%), He (18%)	CH_4, H_2O, NH_3, C_2H_6, PH_3	H_2S, C_2H_2, CH_3D, HCN, CH_3NH_2, N_2H_4, GeH_4, CO
Saturn	H_2 (94%) He (6%)	CH_4, H_2O, NH_3, C_2H_6, PH_3	H_2S, CH_3NH_2, C_2H_2, CH_3D, HCN, N_2H_4, GeH_4, CO
Titan	N_2 (94%), CH_4 (6%)	Ar, H_2, CO, C_2H_6, C_3H_8, C_2H_2	C_2H_4, HCN, CH_3CCH, HC_4H, HC_3N, NCCN, CO_2

Note: Major compounds are defined here as those compounds that have a mole fraction larger than 0.005 in the respective atmosphere, minor compounds as having a mole fraction between 0.005 and 10^{-6}, and trace compounds as having a mole fraction smaller than 10^{-6}.
Minor amounts of methane (about 10 ppb) have been recently detected in the Martian atmosphere by the Mars Express Science Team (Kerr 2004), which could be explained either by life processes, volcanic activity, or geochemical reactions.

cumulate on the planetary surface, as on Titan. Sagan and Salpeter (1976) suggested that life could exist at a level of the Jovian atmosphere where descending organic molecules could be captured and used for energy. The organic-rich atmosphere of Titan, with a density 50% greater than that of the Earth, conceivably could support life in the same way.

Several challenges to life in an atmosphere must be met. A main problem is that the density of gas molecules in an atmosphere is much smaller than on a planetary surface, so to achieve the level of interactions between molecules required for living processes, the density of the organism is inevitably going to exceed that of its surroundings, and buoyancy will be a problem. However, atmospheres can be very dense, achieving liquid-like states at sufficiently high pressures. And just as fishes evolved air bladders to give them buoyancy in water on Earth, an airborne microbe conceivably could evolve a gas-filled organelle that radically increases its volume to a point where its average density is sufficiently low to keep it afloat in the air. Another potential problem could be exposure to high radiation levels, but conditions that would promote the survival of atmospheric organisms would favor the evolution of radiation protection mechanisms (Schulze-Makuch et al. 2004b).

A critical problem for indefinite survival in the atmosphere is the question of a suitable solvent for the support of life-sustaining reactions. Both liquids and solids are generally present in an atmosphere, such as liquid water and aerosols in Earth's clouds. But their abundance in the atmosphere compared to the planetary surface is very low. To be effective solvents, the liquids need to be condensed into droplets of sufficient size and longevity to provide a transiently stable pool of airborne liquid. Such droplets do appear to exist in the upper atmosphere of Venus (Grinspoon 1997), and perhaps in other dense atmospheres elsewhere.

Another potential problem is the scarcity of a solid substratum. The interface of a liquid solvent with a solid surface is presumed to provide a much more likely circumstance for the development of complex chemistry, simply because the degrees of freedom for interacting components are reduced from three dimensions to two. Thus, the origin of life in particular, seems much more likely to come about at interfaces than in three-dimensional volumes of gasses or liquids. Once underway in its confined cellular compartments, life would have an easier time of surviving in three-dimensional volumes, as many organisms in water, and some forms in air, do on Earth. The plausibility of life in an atmosphere is thus higher on those planetary bodies where conditions at the surface were amenable for the origin or early cultivation of life on a solid substrate. This would mean that Venus and Titan, for example, would have experienced a greater chance for the origin of surface (or subsurface) life that eventually evolved adaptations for an airborne existence, than would the gas giant planets, where a solid substrate may never have existed under conditions appropriate for life to originate or take hold. For an alternative view, see Feinberg and Shapiro (1980) who consider that the absence of a surface might be an advantage, because it would allow free motion between different environments, making it possible for an organism to invent its own disequilibrium by moving from one condition to another.

As an instructive example, we will consider the case of Venus in more detail. There is evidence for an early ocean on Venus while the early Sun was fainter than it is now. Life on Earth developed very fast once conditions became appropriate (Chap. 3). The same could have occurred on Venus. Alternatively, life may have been transplanted from Earth or Mars to Venus via meteorite impacts. Either way, life may have become established on Venus at an early point in its history. We know that conditions on the surface of Venus are now inhospitable to life as we know it, with temperatures around 733 K ($\sim 450\,^{\circ}$C) and extreme desiccation. The change in planetary surface conditions was presumably caused by a run-away greenhouse effect as the Venusian atmosphere moved toward its present composition of 97% CO_2. If the environmental transformation occurred slowly enough, microbial life could have adapted to life in the clouds of Venus by directional selection (Schulze-Makuch and Irwin 2002).

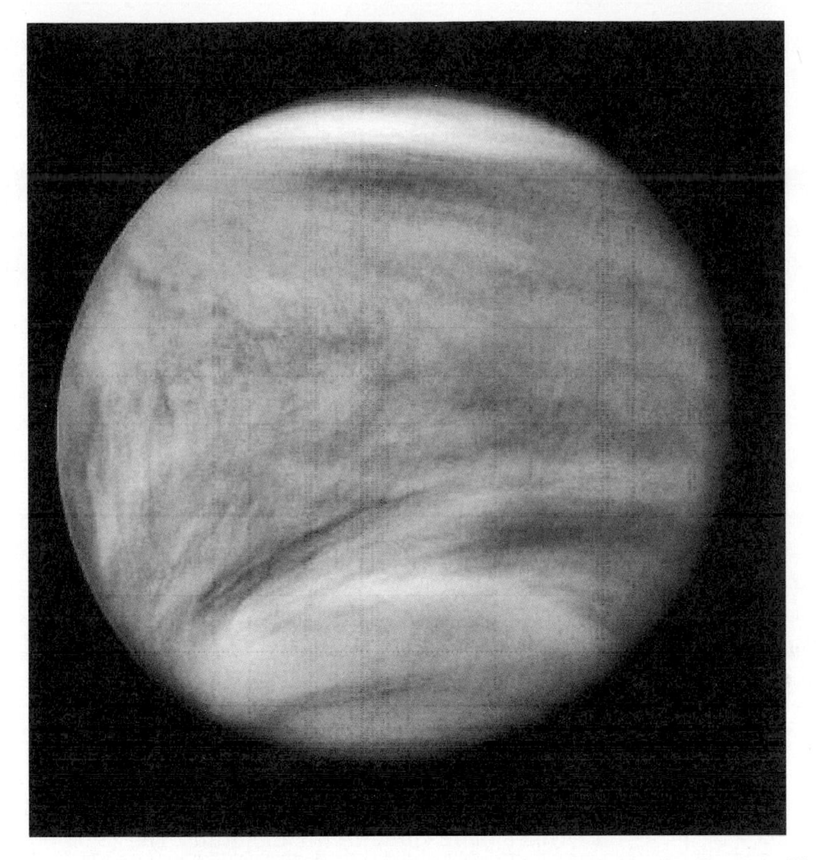

Fig. 7.2 Ultraviolet image of the clouds of Venus as seen by the Pioneer Venus Orbiter (5 February 1979). The dark streaks are produced by absorption of solar UV radiation. Source: NSSDC, http://nssdc.gsfc.nasa.gov/photo_gallery/photogallery-venus.html, image pvo_uv_790205.

Several factors would support such a life style in the atmosphere of Venus: (1) The lower atmosphere is thick, so under liquid-like conditions microbial transport between the surface and the cloud layer would be easier than in Earth's atmosphere. (2) The clouds of Venus are much larger, providing more continuous and stable environments than clouds on Earth. (3) Current conditions in the lower cloud layer of Venus are relatively benign at 300 – 350 K, 1 bar pressure, and a pH of 0 – conditions of temperature, pressure, and pH under which thermoacidophilic microbes are known to thrive on Earth (these are also some of the oldest known forms of life on Earth). (4) Cloud particles are projected to last for several months in the Venusian atmosphere compared to only days on Earth (Grinspoon 1997). (5) The Venusian atmosphere is super-rotating, thus cutting the nighttime significantly and thereby allowing for more photosynthesis. (6) Water vapor is reasonably dense in the lower cloud layers of Venus. (7) Oxygenated species such as SO_2 and O_2 co-exist and are in thermodynamic disequilibrium with reducing species such as H_2S and H_2 in the Venusian atmosphere.

An ultraviolet absorber has been detected in the Venusian atmosphere that may be related to microbial UV protection and possible photosynthesis (Schulze-Makuch et al. 2004b, Fig. 7.2). These and other aspects of the possibility of life in the clouds of Venus have been raised by different authors over the years (Sagan 1961, Feinberg and Shapiro 1980, Grinspoon 1997), and have been discussed more recently in the context of a proposed sample return mission to Venus (Schulze-Makuch and Irwin 2002, Schulze-Makuch et al. 2002). We concluded that Venus provides one of the best possibilities for harboring atmospheric life in the solar system. Since it is also the most accessible planetary body beyond the Moon a sample return mission lies within the capabilities of existing technology.

Venus illustrates nicely the theoretical potential for atmospheric life. Nonetheless, the problems for persistence of living systems in an atmospheric habitat are formidable, so their existence warrants a lower probability than life on the surface, and much lower than life beneath the surface, on other worlds.

7.4 Life in the Space Environment

If the low density of matter would make life-supporting interactions between molecules in a gaseous atmosphere difficult to maintain, the problems are much more severe in space. The damaging potential of ultraviolet and particle radiation, the extremely low temperature and nonexistent vapor pressures, and the homogeneity of empty space further add up to such a hostile environment that outer space can not be regarded as a likely habitat for life. However, the possibility that life could survive interplanetary travel through space in the protective sanctuary of meteorites or even dust particles cannot be discounted. There is increasing evidence that microbes, especially when in the dormant spore form, can survive space conditions fairly well (Horneck 1981, Koike et al. 1991, Nicholson et al. 2000). This is especially the case if the microbe is surrounded by a thin layer of solid material that would shield it from cosmic and UV radiation. The effect of space vacuum is another constrain. Some space experiments have shown that up to 70% of bacterial and fungal spores survive 10 days exposure to space vacuum, even without any protection (Horneck 1993). Survival rates increased when *Bacillus subtilis* spores were embedded in salt crystals or if they were exposed in thick layers (e.g. 30% spore survival after nearly 6 years when embedded in salt crystals; Horneck et al. 1994). Other studies showed that bacterial survival rates decreased by 2 to 4 orders of magnitude when exposed to space vacuum and short wavelength UV radiation (Saffary et al. 2002), but confirmed the protection provided by salt crystals (Mancinelli et al. 1998). An intriguing example of microbial survival under space conditions was the reported recovery of living bacteria from the Surveyor 3 spacecraft after three years of exposure on the lunar surface (Mitchell and Ellis 1971), although this claim has been disputed. Nevertheless, microbes do apparently have the possibility to survive for extended periods in space. *Deinococcus radiodurans* appears to accomplish its resistance to radiation and desiccation by having multiple copies of DNA, large organelles, a large nucleus, a thick membrane, having the DNA in a ring-like structure (Levin-Zaidman et al. 2003), and by possessing a high redundancy of repair genes, but most microbes accomplish this feat by sporulation. During sporulation cytoplasm and genetic material is sealed off by the inner cell membrane. The DNA is then protected by thick layers of protective membranes (Fig. 7.3), which are only permeable to nutrients that the organism needs for germination.

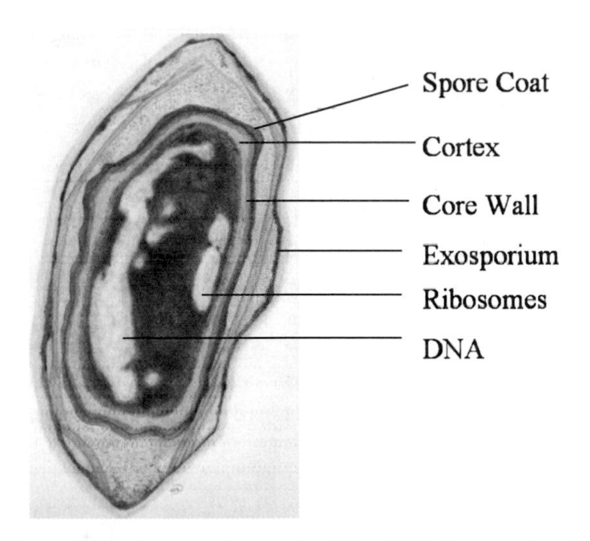

Fig. 7.3 *Bacillus subtilis* spore, schematic. The DNA is contained in the nucleoid (light regions) within the spore core. The core is surrounded by the protective cortex. The long axis of a *B. subtilis* spore is about 1.2 micrometers (Nicholson et al. 2000). Drawing provided by Chris D'Arcy, Dragon Wine Illustrations, El Paso, Texas.

Microbes can survive in this type of dormant phase for an extremely long time. Cano and Borucki (1995) isolated a strain of *Bacillus sphaericus* from an extinct bee trapped in 25–30 million-year-old amber, while Vreeland et al. (2000) claimed to have isolated a 250 million-year-old halotolerant bacterium from a salt crystal.

Findings such as these lend new credibility to the idea of panspermia (Arrhenius 1903, 1908), the transfer of organisms between planetary bodies. However, any organism taking this type of journey would have to survive a series of hazards, including (1) survival of the meteorite impact that ejects the organism into space from the planet of its origin, (2) maintenance of viability for long durations of time inside the meteoritic material, (3) intense UV and cosmic radiation, cold, and vacuum, and (4) the shock and heat of impact on the planetary body to which the organism is transferred. Mileikowsky et al. (2000) and Clark (2001) provided estimates on the likelihood of microbial survival for the different steps. Davies (1996) analyzed this scenario for the Mars-Earth case and concluded that it is a plausible scenario. A critical parameter is travel time, which can be as little as 2 months for microscopic particles from Mars to Earth (Moreno 1988). Boulder-size rocks, however, have been estimated to need a mean travel time of several hundred thousands to millions of years for the same distance (Melosh 1988). Nevertheless, the interplanetary travel from one planetary body in a solar system to another is a definite possibility. However, panspermia between different solar systems is extremely unlikely given the great distances between solar systems and the low statistical probability for a life-transporting meteorite to intersect with a habitable planet.

It should be pointed out that viability in the space environment very likely involves only dormant forms of life. Active forms of life such as speculated by Hoyle (1959, see Sect. 8.2) could not exist due to the harsh radiation environment, cold vacuum conditions and low density, plus the problem of origin. The idea that an ancestor of such an organism would have originated on a planetary surface and later adapted to life in space similar to marine animals and plants that conquered the land during Earth's history, seems unreasonable. There are many transitional habitats between land and sea, but not between a planetary atmosphere and space. Evolutionary pressure would have had to push certain types of organisms to adapt to life in the atmosphere, then pressed it to higher and higher levels of the atmosphere until finally the organism would have to be capable of living in space. Over Earth's history of immense evolutionary pressure during certain time periods, only a tiny fraction of terrestrial organisms adapted even to a life style involving the atmosphere. The major problem appears to be that chemical nutrients that are needed for growth in addition to light are not present in high enough concentrations in the higher atmosphere and certainly not in space.

7.5 Chapter Summary

The human perspective of life as a planetary surface phenomenon is deceiving. The surface provides a heterogeneous environment conducive for the diversification of life over time as conditions change. It is just these circumstances, in all probability, that have given rise to macrobiological complexity on Earth. But the part of the biosphere that lies beneath the surface provides for a more stable and secure abode for life, and may even on Earth harbor a greater total biomass than is found above ground. Thus, life in the subsurface is much more likely to be the rule than the exception on other worlds. If this is so, there are compelling theoretical reasons for believing that in the vast majority of cases, such life is microscopic and relatively ancestral.

The gaseous atmosphere that surrounds planetary bodies is a much less favorable habitat for living systems. But at high densities with an appropriate mixture of chemicals and available free energy, atmospheres could harbor life. Like their subsurface counterparts they would probably be microscopic for reasons having to do with buoyancy, but because of the peculiar evolutionary trajectory that likely led to their adaptation to an aerial existence, they are more likely to be highly derived in form and function from their ancestors. Active life in space is highly improbable due to the harsh radiation environment, cold vacuum conditions and low density, relative homogeneity, and the problem of origin. However, organisms have developed protective mechanisms that allow them to travel passively through space for some time.

References

Arrhenius S (1903) Die Verbreitung des Lebens im Weltenraum. *Umschau* 7: 481–485.

Arrhenius S (1908) *Worlds in the making.* Harper Collins, London.

Boston PJ, Stoker CR (1983) Microbial metabolism of organic molecules produced by chemical synthesis in a reducing atmosphere: implications for the origin of life. In: Pepin RO, O'Connell (eds) *Planetary Volatiles*, Lunar and Planetary Institute, Houston, pp 31–39.

Boynton WV, Feldman WC, Squyres SW, Prettyman T, Brückner J, Evans LG, Reedy RC, Starr R, Arnold JR, Drake DM, Englert PA J, Metzger AE, Mitrofanov I, Trombka JI, d'Uston C, Wänke H, Gasnault O, Hamara DK, Janes DM, Marcialis RL, Maurice S, Mikheeva I, Taylor GJ, Tokar R, Shinohara C (2002). Distribution of hydrogen in the near surface of Mars: evidence for subsurface ice deposits. *Science* 297:81–85.

Cano RJ, Borucki M (1995) Revival and identification of bacterial spores in 25 to 40 million year old Dominican amber. *Science* 268: 1060–1064.

Carr MH (1996) *Water on Mars.* Oxford University Press, Oxford.

Carr MH, Belton MJ, Chapman CR, Davies ME, Geissler P, Greenberg R, McEwen AS, Tufts BR, Greeley R, Sullivan R, Head JW, Pappalardo RT, Klaasen KP, Johnson TV, Kaufman J, Senske D, Moore J, Neukum G, Schubert G, Burns JA, Thomas P, Veverka J (1998) Evidence for a subsurface ocean on Europa. *Nature* 391:363–365.

Chyba CF (1997) Life on other moons. *Nature* 385:201.

Clark B (2001) Planetary interchange of bioactive material: probability factors and implications. *Origins of Life and Evolution of the Biosphere* 31: 185–197.

Coustenis A, Lorenz RD (1999). Titan. In: Weissman M L-A, Johnson TV (eds) *Encyclopedia of the Solar System.* Academic Press, New York, pp 377–404.

Davies PCW (1996) The transfer of viable microorganisms between planets. Ciba Foundation Symposium 202 (Evolution of hydrothermal ecosystems on Earth (and Mars?), Wiley, Chichester.

Dimmick RL, Wolochow H, Chatigny MA (1979) Evidence for more than one division of bacteria within airborne particles. *Applied Environmental Microbiology* 38: 642–643.

Dohm JM, Anderson RC, Baker VR, Ferris JC, Hare TM, Strom RG, Rudd LP, Rice JW, Casavant RR, Scott DH (2000) System of gigantic valleys northwest of Tharsis, Mars; latent catastrophic flooding, northwest watershed, and implications for northern plain ocean. *Geophysical Research Letters* 27: 3559–3562.

Feinberg G, Shapiro R (1980) *Life beyond Earth – the intelligent Earthling's guide to life in the universe.* William Morrow and Company, Inc., New York.

Ferguson BA, Dreisbach TA, Parks CG, Flip GM, Schmitt CL (2003) Coarse-scale population structure of pathogenic Armillaria species in a mixed-conifer forest in the Blue Mountains of northeast Oregon. *Canadian Journal of Forest Research* 33: 612–623.

Friedmann EI (1982) Endolithic microorganisms in the Antarctic cold desert. *Science* 215: 1045–1053.

Fuzzi S (2002) Organic component of aerosols and clouds. EUROTRAC-2 Symposium 2002: Transformation and Chemical Transformation in the Troposphere, Garmisch-Partenkirchen, Germany.

Gislén T (1948) Aerial plankton and its condition of life. *Biological Reviews* 23: 109–126.

Gold T (1992) The deep, hot biosphere. *Proceedings of the National Academy of Sciences (USA)* 89: 6045–6049.

Grinspoon DH (1997) *Venus revealed: a new look below the clouds of our mysterious twin planet.* Perseus Publishing, Cambridge, Massachusetts.

Head JW, Smith D, Zuber M (1998) Mars; assessing evidence for an ancient northern ocean with MOLA data. *Meteoritics & Planetary Science* 33: Suppl. A 66.

Herlihy LJ, Galloway JN, Mills AL (1987) Bacterial utilization of formic and acetic acid in rainwater. *Atmos. Environ.* 21: 2397–2402.

Hoffman PF, Kaufman AJ, Halverson GP, Galen P, Schrag DP (1998) A neoproterozoic snowball Earth. *Science* 281: 1342–1346.

Horneck G (1981) Survival of microorganisms in space: a review. *Adv. Space Research* 1: 39–48.

Horneck G (1993) Responses of Bacillus subtilis spores to the space environment: results from experiments in space. *Origins Life Evol. Biosph.* 23: 37–52.

Horneck G, Buecker H, Reitz G (1994) Long-term survival of bacterial spores in space. *Adv. Space Research* 14: 41–45.

Hose LD, Palmer AN, Palmer MV, Northup DE, Boston PJ, DuChene HR (2000) Microbiology and geochemistry in a hydrogen-sulphide-rich karst environment. *Chemical Geology* 169: 399–423.

Hoyle F (1959) *The black cloud.* Signet, New York.

Imshenetsky AA, Lysenko SV, Kazakov GA (1978) Upper boundary to the biosphere. *Applied Environmental Microbiology* 35: 1–5.

Jannasch HW (1995) Seafloor hydrothermal systems: physical, chemical, biological and geological interactions. Geophysical Monograph 91. American Geophysical Union, Washington DC.

Johnson HP and the LEXEN Scientific Party (2003) Probing for life in the ocean crust with the LEXEN program. *EOS Trans.AGU* 84: 109 & 112.

Kerr RA (2004) Life or volcanic belching on Mars. *Science* 303: 1953.

Khurana KK, Kivelson MG, Stevenson DJ, Schubert G, Russell CT, Walker RJ, Polanskey C (1998) Induced magnetic fields as evidence for subsurface oceans in Europa and Callisto. *Nature* 395: 777–780.

Kirschvink JL, Gaidos EJ, Bertani LE, Beukes NJ, Gutzmer J, Maepa LN, Steinberger RE (2000) Paleoproterozoic snowball Earth; extreme climatic and geochemical global change and its biological consequences. *Proceedings of the National Academy of Sciences (USA)* 97: 1400–1405.

Koike J, Oshima T, Koike KA, Taguchi H, Tanaka R, Nishimura K, Miyaji M (1991) Survival rates of some terrestrial microorganisms under simulated space conditions. *Adv. Space Research* 12: (4)271–(4)274.

Levin-Zaidman S, Englander J, Shimoni E, Sharma AK, Minton KW, Minsky A. (2003) Ringlike structure of the Deinococcus radiodurans genome: A key to radioresistance? *Science* 299: 254–256.

Lewis JS (1995) *Physics and chemistry of the solar system.* Academic Press, San Diego, California.

Malin MC, Edgett KS (2000) Evidence for recent groundwater seepage and surface runoff on Mars. *Science* 288: 2330–2335.

Mancinelli RL, White MR, Rothschild LJ (1998) Biopan survival I: exposure of the osmophiles Synechococcus sp. (Nageli) and Haloarcula sp. to the space environment. *Adv. Space Research* 22: 327–334.

McKinnon W, Kirk RL (1999) Triton. In: Weissman M L-A, Johnson TV (eds). *Encyclopedia of the Solar System.* Academic Press, New York, pp 405–434.

Melim LA, Shinglman KM, Boston PJ, Northup DE, Spilde MN, Queen JM (2001) Evidence for microbial involvement in pool finger precipitation, Hidden Cave, New Mexico. *Geomicrobiology Journal* 18: 311–329.

Melosh HJ (1988) The rocky road to panspermia. *Nature* 332: 687–688.

Mileikowsky C, Cucinotta FA, Wilson JW, Gladman B, Horneck G, Lindegren L, Melosh J, Rickman H, Valtonen M, Zheng JQ (2000) Risks threatening viable transfer of microbes between bodies in our solar system. *Planetary Space and Science* 48: 1107–1115.

Mitchell FJ, Ellis WL (1971) Surveyor III; bacterium isolated from lunar-retrieved TV camera. *Geochimica et Cosmochimica Acta* 2: 2721–2733.

Moreno MA (1988) Microorganism transport from Earth to Mars. *Nature* 336: 209.

Nicholson WL, Munakata N, Horneck G, Melosh HJ, Setlow P (2000) Resistance of Bacillus endospores to extreme terrestrial and extraterrestrial environments. *Microbiology and Molecular Biology Reviews* 64: 548–572.

Onstott TC, Phelps TJ, Colwell FS, Ringelberg D, White DC, Boone DR (1999) Observations pertaining to the origin and ecology of microorganisms recovered from the deep subsurface of Taylorsville Basin, Virginia. *Geomicrobiology Journal* 14: 353–383.

Pedersen K (2000) Exploration of deep intraterrestrial microbial life: current perspectives. *FEMS Microbiology Letters* 185: 9–16.

Pedersen K, Ekendahl S (1990) Distribution and activity of bacteria in deep granitic groundwaters of Southeastern Sweden. *Microbial Ecology* 20: 37–52.

Rzonca B, Schulze-Makuch D (2002) Investigation of hydrothermal sources in the Rio Grande rift region. New Mexico Geological Society Guidebook, 53rd Field Conference, Geology of White Sands: 319–324.

Saffary R, Nandakumar R, Spencer D, Robb FT, Davila JM, Swartz M, Ofman L, Thomas RJ, DiRuggiero J (2002) Microbial survival of space vacuum and extreme ultraviolet irradiation: strain isolation and analysis during a rocket flight. *FEMS Microbiology Letters* 215: 163–168.

Sagan C (1961) The planet Venus. *Science* 133: 849–858.

Sagan C, Salpeter EE (1976) Particles, environments, and possible ecologies in the jovian atmosphere. *Astrophys. J. Suppl. Ser.* 32: 624.

Sattler B, Puxbaum H, Psenner R (2001) Bacterial growth in supercooled cloud droplets. *Geophysical Research Letters* 28: 239–242.

Schlesinger WH (1997) *Biogeochemistry*, 2nd edition. Academic Press, New York, USA.

Schulze-Makuch D (2003) Chemical and microbial composition of subsurface-, surface-, and atmospheric water samples in the southern Sacramento Mountains, New Mexico. Proceedings of the New Mexico Geological Society Annual Spring Meeting, Socorro, New Mexico, pp 62.

Schulze-Makuch D, Irwin LN (2002) Reassessing the possibility of life on Venus: proposal for an astrobiology mision. *Astrobiology* 2: 197–202.

Schulze-Makuch D, Dohm JM, Fairén AG, Baker VG, Fink W, Strom RG (2004a) Comparative planetology of the inner planets of the solar system: geologic setting, astrobiological assessment and implications for mission design. In review at *Astrobiology*.

Schulze-Makuch D, Grinspoon DH, Abbas O, Irwin LN, Bullock MA (2004b) A sulfur-based survival strategy for putative phototrophic life in the Venusian atmosphere. *Astrobiology* 4: 11–18.

Schulze-Makuch D, Irwin LN, Irwin T (2002) Astrobiological relevance and feasibility of a sample collection mission to the atmosphere of Venus. Proceedings of the 2nd European Workshop on Exo-Astrobiology, ESA-Sp. 518: 247–252.

Showstack R (1998) Lunar prospector finds signature for water ice on Moon, NASA announces. *EOS, Trans. Am. Geophys. Union* 79: 138&144

Slade (1992) Mercury radar imaging: Evidence for polar ice. *Science* 258:635–640

Sleep NH, Zahnle K (1999) Vestiges of living at ground zero. Geological Society of America annual meeting. Abstracts with Programs – Geological Society of America 31: 239–240.

Stoker CR, Boston PJ, Mancinelli RR, Segal W, Khare BN, Sagan C (1990) Microbial metabolism of tholin. *Icarus* 85: 241–256.

Sun HJ, Friedmann EI (1999) Growth on geological time scales in the Antarctic cryptoendolithic microbial community. *Geomicrobiology Journal* 16: 193–202.

Vreeland RH, Rosenzweig WD, Powers DW (2000) Isolation of a 250 million-year-old halotolerant bacterium from a primary salt crystal. *Nature* 407: 897–900.

Whitman WB, Coleman DC, Wiebe WJ (1998) Prokaryotes: the unseen majority. *Proceedings of the National Academy of Sciences (USA)* 95: 6578–6583.

8 Ideas of Exotic Forms of Life

Science and speculation have converged at the boundaries of human imagination to conceive of some very exotic states of matter and/or energy that have been claimed by their authors to represent alternative forms of life or to exhibit life-like characteristics. Those ideas have been advanced by serious thinkers and thus deserve to be evaluated in the context of our assumptions about the fundamental nature of life. We will briefly mention some of the most important ideas proposed and critically examine them in light of our proposed definition advanced in Chap. 2 that life is (1) composed of bounded microenvironments in thermodynamic disequilibrium with their external environment, (2) capable of transforming energy and the environment to maintain a low-entropy state, and (3) capable of information encoding and transmission. Only if all of those three criteria are met does the proposed idea constitute a viable alternative form of life in our view.

8.1 Life Based on Spin Configurations

In Sect. 4.2.9 the possibility was discussed of obtaining energy from spin configurations. Feinberg and Shapiro (1980) took this idea a step further and speculated on the possibility of life based on spin configurations of p-hydrogen and o-hydrogen (Fig. 4.7). They suggested an organism with a helium interior, an inner o-hydrogen layer and an outer p-hydrogen layer, which would be capable of controlled processes and obtaining energy by the process described in Sect. 4.2.9. The living environment envisioned was a very cold and dark planet, just a few tens of degrees above absolute zero. In this type of environment solid hydrogen would be floating in a sea of liquid hydrogen. An input of energy, most suitably from a faint star with emissions in the weak infrared and microwave wavelengths, would be absorbed by hydrogen, thereby transfering some of the hydrogen into the energy-rich ortho-hydrogen state. The energy stored in the o-hydrogen state would then be released when the atoms are transformed into the p-hydrogen state. Feinberg and Shapiro suggested that for such life to form, a precise arrangement of o- and p-hydrogen would need to be established.

However, it is difficult to envision how such an organism could perform work efficiently without chemical reactions. In addition, the force that would hold the unbounded organism together against the tendency toward entropy is unclear. Specifically the inner energy-rich o-hydrogen molecules would have to be kept apart from each other to avoid catalyzing their own destruction. The function of the helium interior is unclear as well.

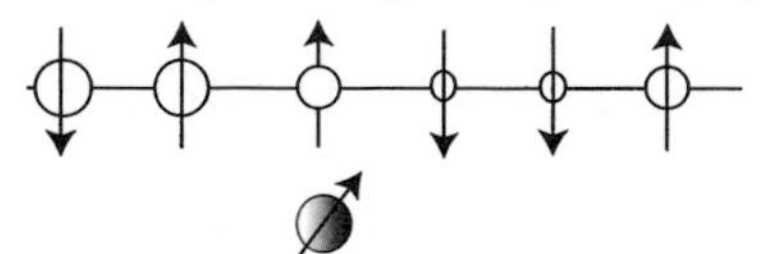

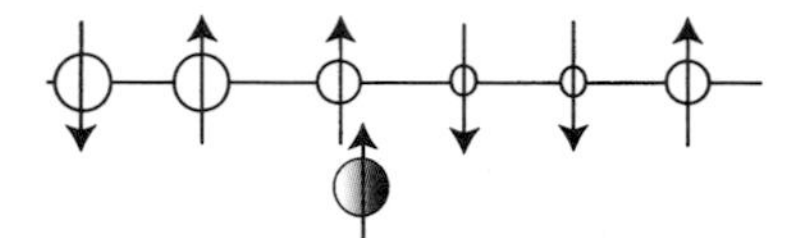

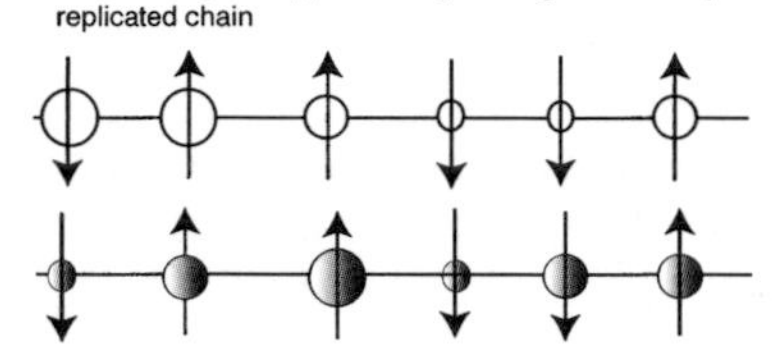

Fig. 8.1 Hypothetical replication mechanism of an ordered chain of magnets. Notice that identical components are replicated rather than complimentary and that the chain of magnets is only sensitive to the direction of magnetization, not to magnet size (modified from Feinberg and Shapiro 1980).

Feinberg and Shapiro (1980) do address the issue of replication (though not in direct association with the proposed organism). Information encoding and transmission would not need to occur chemically, but rather could be based on magnetic orientation (Fig. 8.1). The starting point would be a chain of atomic magnets with their magnetic moments aligned in variable directions. A randomly directed magnet approaching the chain would line up with its direction parallel to that of the nearest magnet. If that process were continued for many magnets, a new chain would eventually be formed that duplicates the original chain in the directional arrangement of its magnets. If the magnets in place along the chain retained their alignment and could be protected from re-magnetization from an exterior field, such an informational string would be a realistic possibility for replication and transmission of biological information. However, for such a code to be consistent with our definition of life, the code-bearing material would need to be distinct from the living entity that harbors it. Such a distinction is not made, nor is it clear how the information in such a magnetic string would be transformed into physical or chemical operations associated with living processes.

Thus, while this is a stimulating and intriguing idea for an exotic form of life, especially under conditions of extreme cold, which are common in the universe, we consider the existence of such an organism as extremely unlikely.

8.2 Fred Hoyle's Black Cloud and Similar Ideas

The Black Cloud organism envisioned by Hoyle (1959) is portrayed as a formless mass living in space, obtaining energy from the light of stars, and communicating with itself and other forms of life by radio waves. It has an organization analogous to that of complex terrestrial animals, such as gas that has the function of blood, an electromagnetic heart and kidneys, and a complex neurological system that can be understood as a brain. In his novel Fred Hoyle entertains the fiction that a gaseous being in the universe would have definite advantages compared to organisms living on a planetary surface, because (1) it would be free of constraint from large gravitational forces, and (2) it would be able to absorb much more energy from starlight than the minute amount harvested by organisms confined to a planetary surface.

An organism like this is imaginable in principle (Goldsmith and Owen 2001). However, the major problem with this idea is the difficulty of envisioning how it could arise from an inanimate origin in the space environment. Being unbounded, the "proto-organismic" matter would have a low density, and be exposed to cosmic radiation. Low density and radiation would counteract any tendency for the matter within such an organism to become organized. Furthermore, no mechanism of replication is envisioned.

A somewhat similar concept was proposed by Arvidas Tamulis and coauthors (e.g. Tamulis et al. 2001, 2003), in the form of a molecular quantum computing cloud that could absorb magnetic and light energy from planets and stars, compute information, and move in space by using light pressure. Tamulis et al. (2003) pointed out the similarity to molecular quantum computers that use photoactive molecules converting light energy to magnetic flops interacting and controlling the central generating element of 10 quantum bits. An essential requirement for quantum computing life would be long lasting coherent quantum states, which would only be possible at extremely low temperatures found in interstellar dust clouds and on very cold planets. Basically, this idea suffers from the same problems as Hoyle's Black Cloud. An intriguing idea, but it lacks key fundamental characteristics of living systems.

8.3 Life on a Neutron Star

Life based on the strong interaction (strong nuclear force) rather than electromagnetic energy has been suggested as a possible basis for life as well (Feinberg and Shapiro 1980, Goldsmith and Owen 2001). The strong interaction affecting quarks, antiquarks, and gluons (carrier particle of the strong interaction) holds together the nuclei of atoms, but it is a force that is only strong at extremely short distances. Thus, if atoms are in their (for us) usual state with electrons orbiting the nucleus, the individual nuclei are too far apart for the strong atomic force to result in significant interactions between different nuclei. However, given the situation that electrons are ripped off from their nuclei, protons and neutrons from various nuclei come into close proximity. This requires immense gravitational forces, and occurs naturally in neutron stars, which have one to two solar masses concentrated in a body with a diameter of 10 to 20 km and a magnetic field of about 10^{12} gauss. Given the high density and temperatures, interactions between protons and neutrons occur much more frequently than electromagnetic force interactions. These interactions occur because of incredible gravitational and magnetic forces present on the surface of a neutron star. Ruderman (1974) pointed out that the magnetic forces would reshape the "normal" atoms into strange configurations exhibiting long polymer-like chains in which the nuclei lie along a central line and the electrons in elongated bands. Magnetically formed polymers could then even align to form larger structures.

Life on a neutron star based on the strong interaction is one of the most extreme applications of energy in a conceivably biological context. The idea provides some conceptualization of how dynamic complexity can be established, but falls far short of constituting a comprehensive model of a living system, including any meaningful definition of life such as the one offered in Chap. 2. Neutron stars, however, might have planets, and possible life on such a planet could use the extremely strong magnetic field of their star as a primary energy source (see Sect. 4.2.5). In fact, the first extrasolar planets were discovered around a neutron star, the pulsar PSR 1257+12 (Wolszczan and Frail 1992; see also Sect. 9.4).

8.4 Life on a Brown Dwarf

Brown dwarfs are accumulations of gas that have not been able to increase their mass and temperature sufficiently to sustain hydrogen fusion and become a star. Low-mass brown dwarfs may contain liquid water, and could possibly be capable of supporting life (Shapley 1958). Energy on a brown dwarf could be provided from the body's own intrinsic infrared spectrum, though this form of energy is much weaker than light in the visible and near-infrared spectrum. If life could adapt to use the low infrared spectrum to obtain energy instead of the visible and near-infrared spectrum as used on Earth, organisms relying on photosynthesis would have a nearly inexhaustible supply of energy. Such organisms would have to adapt to the strong gravitational field of brown dwarfs, perhaps 100 times stronger than on Earth. However, that may not be a problem. Mastrapa et al. (2001) tested *Deinococcus radiodurans* and *Bacillus subtilis spores* by exposing them to extreme acceleration (4.5×10^6 m/s^2) and jerks (1.5×10^{11} m/s^3) in a compressed-air pellet rifle and noted survival rates between 40 and 100%. Also, Sharma et al. (2002) observed physiological and metabolic activity of *Shewanella oneidensis* strain MR1 and *Escherichia coli* strain MG1655 at pressures of 68 to 1680 MPa in diamond anvil cells. However, other major problems would include the lack of available surfaces for chemical reactions, suitable temperatures, and the relative lack of heavier elements such as potassium, calcium, and iron, which are necessary for living processes with which we are familiar. A more conventional possibility would be life on planets that are orbiting brown dwarfs. Andreyeschchev and Scalo (2002), for example, modeled habitable distances and time scales for planets orbiting brown dwarfs. Near-infrared observations indicate that young brown dwarfs probably possess proto-planetary disks (Muench et al. 2001, Testi et al. 2002), thus life could conceivably exist on a planet in a brown dwarf system.

8.5 Some Other Ideas on Forms of Exotic Life

Some other ideas of exotic life have been proposed in the past. They include speculations of plasma life inside a star, which was first advanced by Maude (1963) and later elaborated on by Feinberg and Shapiro (1980). The idea is based on replication of certain patterns of magnetic force and the dynamic activities within a star. We consider these possible interactions as interesting examples of dynamic complexity arising from relative chaos similar to forest fires and hurricanes on Earth, but do not consider them alive by any meaningful definition of the term. Other speculations of exotic biology include life based on pure energy and life based on topological effects in quantum space rather than atoms as suggested in a novel by Egan (2002). As intriguing as these speculations are, any ideas not involving atomic matter are farfetched and have to remain in the realm of science fiction for now. Further, unlike matter, where two atoms or two planets affect each other's behavior through collisions, two independent flows of radiation will usually pass through each other without having any significant effect on each other. Thus, these ideas are imaginative examples of exotic physics, but, in our view, do not have a recognizable relationship to any meaningful definition of life.

8.6 Chapter Summary

We have given these brief examples of other, more exotic notions about forms of life that transcend even the basic chemical and physical laws that operate within usual planetary dimensions, because they have been advanced by serious thinkers who force us to critically examine our assumptions about the fundamental nature of life. In no case, however, do we feel that any of these examples constitute a plausible argument for an alternative form of life. They embody some imaginative alterations of state, matter, and energy, and bear some resemblance to some characteristics of life – particularly those having to do with the organizing capacity of energy flow and the tendency to create low-entropy states of disequilibrium. But in their totality, they do not come close to meeting the criteria for living systems set forth in Chap. 2. They do not prescribe bounded environments, nor specify a mechanism for reproduction, nor identify informational storage mechanisms that persist in unitary form from one generation to another. It perhaps could be argued that the flaw in our analysis lies in the definition of life, rather than its application. In other words, perhaps we lack the imagination to envision all the possibilities by which living systems could be manifested. However, we believe that in this book we have already pushed the limits of what it means to be alive, and to push beyond those limits at this point in our understanding would take us into the realm of a speculative physics that has no recognizable relationship to life in the universe.

References

Andreyeschchev A, Scalo J (2002) Duration of habitability of brown dwarf planets. *Bioastronomy* 2002: Great Barrier Reef Conference Proceedings, in press.

Egan G (2002) *Schild's Ladder.* Eos, New York.

Feinberg G, Shapiro R (1980) *Life beyond Earth – the intelligent Earthling's guide to life in the universe.* William Morrow and Company, Inc., New York.

Goldsmith D, Owen T (2001) *The search for life in the universe.* Benjamin/Cummings Publishing Company, Menlo Park, California.

Hoyle F (1959) *The black cloud.* Signet, New York.

Mastrapa RME, Glanzberg H, Head JN, Melosh HJ, Nicholson WL (2001) Survival of bacteria exposed to extreme acceleration; implications for panspermia. *Earth and Planetary Science Letters* 189: 1–8.

Maude AD (1963) Life in the Sun In: Good IJ (ed) *The Scientist Speculates,* Basic Books, Inc., New York.

Muench AA, Alves J, Lada CJ, Lada EA (2001) Evidence for circumstellar disks around young brown dwarfs in the Trapezium cluster. *Astrophys. J.* 558: L51–54.

Ruderman M (1974) *Physics of dense matter.* Hansen C (ed), D. Reidel Publishing Company, Dordrecht, The Netherlands.

Shapley H (1958) *Of stars and men.* Beacon Press, Boston.

Sharma A, Scott JH, Cody GD, Fogel ML, Hazen RM, Hemley RJ, Huntress WT (2002) Microbial activity at gigapascal pressures. *Science* 295: 1514–1516.

Tamulis A, Tamuliene J, Balevicius ML, Rinkevicius Z (2001) Ab initio quantum chemical search of per linear transition state of azo-dye molecules and design of molecular logical machines. *Nonlinear Optics* 27: 481–488.

Tamulis A, Tamuliene J, Tamulis V, Ziriakoviene A (2003) Quantum mechanical design of molecular computer elements suitable for self-assembling to quantum computing living systems. Proceedings of the 6th International Conference on Self-Formation, Theory and Applications, 26–28 November, Vilnius, Lithuania; in press at *Solid State Phenomena.*

Testi L, Natta A, Oliva E, D'Antona F, Comeron F, Baffa C, Comoretto G, Gennari S (2002) A young very low mass object surrounded by warm dust. *Astrophys. J.* 571: L155–159.

Wolszczan A, Frail DA (1992) A planetary system around the millisecond pulsar PSR 1257+12. *Nature* 355: 145–147.

9 Signatures of Life and the Question of Detection

In this chapter we will elaborate on how evidence for life on other worlds can be sought, and if present, possibly detected. The best evidence for extraterrestrial life, of course, would be recovery of actual specimens or their fossils. For the next one or two decades, the possibility of obtaining such direct evidence is almost surely restricted to samples from Mars and Venus. So detection of life beyond our nearest neighbors will be dependent for the near future on remote sensing. As technology of robotic exploration and remote sensing improves, the possibility of detecting extraterrestrial life will grow. While the size of individual organisms makes their detection at a distance virtually impossible, organisms in the aggregate alter their environments, generating signatures of their functional processes. These direct consequences of biological activity are referred to as "biosignatures". Other effects of the presence of living systems may be detected in global or geological features. These alterations of the geological environment due to life processes, we call "geosignatures". Even on worlds too remote, small, or difficult for whatever reason to monitor for the existence of explicit signatures of life, certain planetary characteristics can be detected that are more likely to be consistent with the presence of life than others. These we refer to as "geoindicators". They consist of parameters that are consistent with life as defined in Chap. 2, and the requirements for life as described in subsequent chapters, including a flow or gradient of energy, presence of an appropriate solvent, and availability of complex polymeric chemistry. While geoindicators point to the potential for supporting life, they do not confirm its existence. Most geoindicators can be detected by remote sensing methods with relative ease, however, and thus can be used in assessing the plausibility of the existence of life. At the end of the chapter, we apply our discussion of signatures and indicators of life to assess the relative plausibility for the existence of life on other bodies in our solar system, and discuss recent results on extrasolar planetary detection and their implications for astrobiology.

9.1 Biosignatures and Geosignatures of Life

Biosignatures and geosignatures of life are currently not known to exist from any planetary body of our solar system other than Earth. Examples of signatures of life are given in Table 9.1. Thus, there is at present no available evidence for life as we know it elsewhere in the solar system. Life could exist nonetheless, either in a form known or unknown to us, that does not give rise to any of the biosignatures or geosignatures indicated in Table 9.1, if it (1) occurs beneath an opaque surface, (2) is too small to cause environmental transformations extensive in magnitude or spatial extent, or (3) is insufficiently complex to generate complex phenomena, such as roads or radiowaves. Other difficulties are that extraterrestrial life may involve dynamic processes that occur on (1) a spatial scale too small to be detected by current remote technology, and (2) a time scale too prolonged to be sampled feasibly (Schulze-Makuch et al. 2002a). Nevertheless, there are several signatures of life that may become relevant for the detection of life in the near future, either for life in our solar system or in another solar system. These signatures are discussed below.

Table 9.1 Some examples of biosignatures and geosignatures of life.

Observation	Signature
Organic macromolecules larger than 1000 atomic mass units (amu)	Biosignature
Atmospheric gas composition, such as O_2 and CH_4, resulting from biogenic processes	Geosignature
Rocks and sediments produced by biogenic processes such as the banded-iron formation (BIF) and stromatolite deposits of early Earth	Biosignature
Known biogenic substances such as chlorophyll not explicable by naturally occurring inorganic chemical processes	Biosignature
Rate and type of erosion consistent with biological processes	Geosignature
Structural complexity, such as geometric regularity (roads, canals) or unnatural local aggregates (insect colonies, cities) not explicable by natural geological processes	Geosignature
Distribution and magnitude of emitted heat inconsistent with an abiotic origin	Biosignature
Energetic emissions such as radiowaves, which are neither highly regular, as from a pulsar, nor highly random, as in the universal background radiation	Biosignature

9.1.1 Atmospheric Composition of a Planetary Body

An often-cited geosignature of life is the presence of molecular oxygen and particularly the presence of ozone in an atmosphere. For example, Akasofu (1999) suggested the use of the green oxygen line at 557.7 nm from auroral emissions to search for extraterrestrial life. Ozone was suggested to be more suitable than molecular oxygen because its abundance increases nonlinearly with the abundance of molecular oxygen (Leger et al. 1993) and ozone absorbs UV radiation known to be detrimental for terrestrial life on extrasolar planets. Any such spectroscopic remote observation has two major technical challenges: the weak signal and the huge background from the parent star (Frey and Lummerzheim 2002). However, for the detection of the habitability of planets in other solar systems this may be the only reasonable approach for the near future. In our view, an oxygen atmosphere or ozone layer alone, in the absence of other abnormal concentrations of gases such as methane, should be regarded only as a geoindicator consistent with the presence of life, not as a geosignature. For example, Jupiter's moon Europa currently has a thin oxygen atmosphere from interactions of radiation with surface ice (Hall et al. 1995) and it can easily be envisioned that it had a much thicker oxygen atmosphere and possibly an ozone layer for part of its geological history. When the solar system formed Europa most likely had oceans of water on its surface and a water vapor atmosphere. Given the high radiation environment of the Jovian system, water would have split into hydrogen and oxygen with the hydrogen escaping to space and the oxygen being retained for longer time periods because of its higher molecular weight. Thus, the presence of a high amount of molecular oxygen and even ozone for some time period is absolutely plausible based on physical means alone (Europa likely experienced global re-melting events for which the above scenario may be valid as well).

However, the atmospheric composition of Earth is a prime example of a signature of life (Table 9.1). Earth's atmosphere is the peculiar product of a particular biological process: photosynthesis. Oxygen by itself could not be considered a signature, but the high amounts of oxygen (ca. 21%) combined with the presence of hydrogen (H_2), methane (CH_4), ammonia (NH_3), methyl chloride (CH_3Cl), and methyl iodide (CH_3I), along with various sulfur gases can best be explained by the continuous metabolic production of these compounds faster than they can react with each other (Sagan 1994). These gases, highly reactive when mixed, would not coexist at such high concentrations unless their levels were being actively maintained. It is this type of disequilibrium, in combination with high amounts of oxygen, that can be used as an indicator for oxygen-producing photoautotrophs.

9.1.2 Geological Evidence

Particular signatures also exist for chemoautotrophic organisms. An example is provided by the limestones and ironstones produced by biological activity on the early Earth. Both types of rocks can form from inorganic processes. The large quantities produced during Earth's early history, however, can hardly be explained by abiotic processes. Chemosynthesis generates various chemical end-products depending on the exact metabolic process. Nevertheless, the chemical end-product may provide a useful marker, especially if produced in a large enough amount over an extended period of time to make it a signature of chemotrophic life. The biochemical end products often exhibit large-scale geomorphological characteristics such as stromatolite colonies and coral reefs, some of them large enough to be observed with the naked eye from the Moon – such as the Great Barrier Reef.

The high rates of erosion and types of erosion observed on Earth due to biological and chemical weathering induced by living organisms provide another example of a geosignature. The biomass of fungus-lichen rock dwellers is estimated to be enormous, by one account 13×10^{13} tons (Margulis 1998). Thus, the effect of these rock dwellers on chemical weathering from metabolic by-products is immense. Rates and types of erosion can be inferred from the visible and microwave wavelengths of the electromagnetic spectrum (Schulze-Makuch et al. 2002a), but are traditionally not considered as a signature of life. The reason may be that high-resolution images are necessary for unequivocal detection, and this type of resolution is only available for the inner planets of our solar system. It would not be expected to be available for any extra-solar-system planets, particularly not for any encased by an atmosphere.

9.1.3 Fossil Evidence

Fossil remnants and isotopic fractionation caused by biotic processes are other examples of biosignatures. One controversial example is the Martian meteorite ALH84001 in which McKay et al. (1996) claimed to have found evidence of fossilized microbes. However, these claims have come under intense scrutiny, as have some of the oldest known records of life such as stromatolites and microfossils (Pasteris et al. 2002). Another form of fossil evidence are isotopic signatures of carbon, sulfur, nitrogen, hydrogen, iron, and other elements (Schidlowski et al. 1983). Observations from terrestrial organisms show that chemically lighter isotopes are preferred, resulting in a net fractionation of lighter isotopes. A carbon isotope fractionation, typical for biological processes, has been found in the geological record for the last 3.5–3.8 billion years of Earth's history (Schidlowski 1988).

A robotic mission would be needed to detect these signatures of life unless a rare fortunate circumstance happened to bring a meteorite from that world to Earth where it could be analyzed by *in-situ* methods. However, microbial biofilms that become preserved on rock surfaces could possibly be identified with remote sensing methods if (1) spectroscopically identifiable compounds exist that display unique adsorption, diffraction, and reflection patterns characteristic of biogenerated organic compounds (e.g. chlorophylls, carotenes, melanins), (2) biogenic geomorphological features are exhibited (e.g. biopitting, biochipping, bioexfoliation), and (3) biominerals are detected that are produced in association with biofilms that occupy rock surfaces such as oxalates and certain types of carbonates and sulfides (Gorbushina et al. 2002).

9.1.4 Macromolecules and Chirality

Among the most powerful biosignatures are macromolecules that are directly linked to biogenic metabolism or other cellular functions. Chlorophyll is the prime example and can be identified by radiance spectra in the visible region (Hovis et al. 1980, Gordon et al. 1980) and by advanced very-high-resolution radiometer (AVHRR) measurements (Gervin et al. 1985, Tucker et al. 1985). Proteins, polypeptides, and phospholipids are other examples of macromolecules that are linked to life. In general, any macromolecule of a size larger than 1000 amu (protein-size) can be considered a biosignature (Table 9.1).

Chirality, or non-racemic handedness, is a fundamental property of terrestrial biogenic molecules and thus may be used as an indicator of possible extraterrestrial life detectable by remote sensing in the near future. Large macromolecules are not symmetrical and thus inevitably exhibit chirality. Plaxco and Allen (2002) pointed out that all terrestrial life uses well-structured, chiral, stereo-chemically pure macromolecules of 500 or more atoms as their metabolic catalysts. Xu et al. (2003) argued that all life would employ these types of macromolecules irrespective of the specifics of their chemistry. They pointed out that these molecules strongly absorb at terahertz frequencies and exhibit significant circular dichroism, which they consider an unambiguous biosignature. Left- and right-handed circularly polarized light interacts differentially with chiral molecules, especially at blue-green and shorter wavelengths (Van Holde et al. 1998). Salzman et al. (1982) found that organisms also scatter circularly polarized light differentially, with angular and wavelength spectra somewhat characteristic of particular organisms or strains. Evidence presented by these authors as well as Nicolini et al. (1991), Diaspro et al. (1991) and Lofftus et al. (1992) indicates that the configuration of DNA in organisms strongly affects their differential scattering. This indicates that macromolecules and their resulting chirality have a great potential as useful biosignatures.

9.1.5 Presence of Metabolic By-Products and End-Products

Metabolic by-products and end-products are well known for organisms on Earth. They include various biochemical compounds such as ATP and lipids, but also electron donor and acceptor pairs such as Fe^{3+}/Fe^{2+}, NH_3/N_2, H_2S/S enriched in lighter isotopes. This isotope enrichment or fractionation occurs as part of the metabolic reactions for organisms on Earth and may also occur for life elsewhere. While biochemical macromolecules such as ATP are very specific signatures for certain biological processes, using isotopically light electron donors or acceptors as signatures for life is more challenging. There are an endless number of possible electron donor/acceptor pairs that could potentially be used for energy-harvesting reactions on other worlds, and there are also numerous inorganic processes that lead to isotopic fractionation, many of them poorly understood. Nevertheless, the presence, especially of gaseous electron acceptors and donors (e.g. H_2S, COS, CH_4), enriched in lighter isotopes may constitute a signature of life and can be screened readily by remote sensing methods. One example is the presence of CH_4 on Titan, which is isotopically lighter than would be expected from Titan formation theory (Lunine et al. 1999). This was interpreted by Schulze-Makuch and Abbas (2004) and Schulze-Makuch (2004) as a possible signature of life, conceivably resulting from an energy-yielding metabolic reaction of acetylene and hydrogen gas to methane. However, any such interpretation is limited by our understanding of the physical and chemical processes occurring on a planetary body as foreign to us as Titan.

9.1.6 Production of Biogenic Heat

Another possible signature of life is biogenic heat that may be detectable in the future by more advanced technologies. Living systems exist in thermodynamic disequilibrium by drawing energy from their environments. A consequence of the biochemical reactions that an organism needs to carry out to sustain itself is the production of "unorganized energy", commonly in the form of heat. The production of heat follows as a consequence of the second law of thermodynamics. Organisms by their very nature have to be structured and organized. However, in order to conform to the tendency of the physical world toward a state of greater disorder, any organism has to give up a portion of its energy in the form of heat or another type of disorganized energy. The distribution and magnitude of heat produced by living systems or colonies of living systems may be possible to detect by *in-situ* monitoring or remote sensing techniques in the near future, thereby serving as a biosignature if an abiotic origin can be ruled out.

9.1.7 Signatures of More Advanced Life

Signatures of life also include structural complexity produced by biogenic processes ranging from termite mounds to artificial constructions such as streets and evidence of an advanced civilization such as a Dyson Sphere (a constructed spherical shell centered around a star absorbing most of its visible and shorter wavelength radiation; Dyson 1959), that would imply the presence of designed, dynamic activity known only to living systems. Energetic emissions, such as radiowaves, which are neither highly regular, as from a pulsar, nor highly random, as in the universal background radiation, are currently used by SETI (Search for Extraterrestrial Intelligence) to scan the skies for signs of extraterrestrial intelligence. These kinds of signatures, of course, would be linked directly to the presence of more technologically advanced forms of life than microbes, but, if present, would also imply the presence of microbial life based on the presumption that more complex organisms would have to evolve from simpler ancestors.

9.2 Geoindicators of Life

With the exception of Mars and Venus, the detection of life elsewhere in the universe for the foreseeable future has to focus on remote detection given our current state of technology (Schulze-Makuch et al. 2002a). Retrieval of samples for direct analysis for years to come will be limited to meteorites, comet material returned to Earth by the Stardust Mission in 2007, and Venusian atmospheric samples and Martian surface samples possibly retrievable within the next decade or two. Thus, for the foreseeable future it is not feasible practically and economically to send robotic landers to each planetary body of our solar system and beyond. Further, even on a planetary body with life the detection may be unsuccessful if (1) the site is unsuitable or sparsely populated, or (2) life detection experiments are not set up appropriately to detect life thriving in that particular environment.

Under these circumstances, the more sophisticated and abstract definitions of life alluded to in Chap. 2 may provide the basis for a set of parameters that could point to conditions favorable for generic forms of life, either known or unknown to us. Specifically, (1) the maintenance of disequilibrium from the environment requires the availability of energy flow (hence gradients of energy) for sustaining low entropy states; (2) the level of chemical complexity required to transform and store energy appears to require a fluid medium where concentrations can be high but molecular mobility can be maintained; and (3) the storage and transmission of information appears to require polymeric chemistry that can involve the making and breaking of covalent bonds with relative ease. Parameters that indicate the presence of any of these conditions, and therefore imply that life could be present though not confirming its existence, are defined here as geoindicators. Based on the foregoing discussion primary geoindicators of life would include evidence of (a) an atmosphere or ice shield, (b) thermal gradients and chemical disequilibrium conditions, (c) internal differentiation of the planetary body, implying the capacity for radiogenic heating, (d) complex polymeric chemistry, (e) energy flow or gradients, and (f) a liquid medium as a solvent. The advantage of these geoindicators is that remote sensing can readily detect all of them in principle.

9.2.1 Presence of an Atmosphere or Ice Shield

It is difficult to envision the presence of life on the surface of any planetary body that is not shielded by an atmosphere. Without an atmosphere any liquid or gaseous compound will vaporize into the vacuum of space. Aside from the gas giants, relatively dense atmospheres exist only on Earth (1 atm), Venus (~ 90 atm) and Titan (~ 1.5 atm). However, for life to thrive on the planetary surface, temperatures on Venus (very hot) and Titan (very cold) would require a biochemistry with properties unfamiliar to life forms on Earth. The surfaces of both Venus and Titan are obscured from visual light penetration by a thick atmosphere. On Titan, organic compounds such as methane and ethane are present in the atmosphere (Lorenz 1993, Coustenis and Lorenz 1999), and Titan is also the only planetary body with a significant atmosphere other than Earth known to have nitrogen as the most abundant atmospheric gas. The high nitrogen content in Earth's atmosphere has been interpreted to result from biological processes (Lovejoy 2000). Since Titan's atmosphere can only be penetrated by narrow frequency windows between bands for methane and radar with current remote sensing technology (Griffith et al. 1991, Lorenz and Lunine 1997), probes have to be sent to explore the physical conditions and chemistry of the surface (e.g. the Huygens probe scheduled to descend through Titan's atmosphere in January of 2005). On Titan a warmer subsurface could be a suitable habitat for microbial life. Venus, on the other hand, would provide very little hospitality for microbes in the subsurface habitat, unless they were able to use water that might possibly be present in a supercritical state. Instead, if life evolved on Venus, it may have retreated toward cooler conditions in the atmosphere (Schulze-Makuch and Irwin 2002b). Mars has a much thinner but still significant atmosphere dominated by CO_2. The Martian atmosphere would not provide much protection for any life on its surface, but life would be possible in protected niches such as caves or beneath the surface (Boston et al. 1992).

The Jovian moons Europa and Ganymede, and possibly Callisto as well as Neptune's moon Triton, do not have significant atmospheres, but suitable conditions for life in a planetary ocean, if it exists, would be protected by an ice crust. This ice crust would act as a shield preventing subsurface compounds from evaporating into space and would also provide a shield against cosmic rays. Planetary oceans capped by an ice shield may in fact be much more common in the universe than "naked" or "near-naked" oceans as on Earth (Schulze-Makuch 2002).

9.2.2 Internal Differentiation

Life is easier to envision on any planetary body that is differentiated into a radioactive core, a mantel, and a crust. Internal differentiation is a sign of endogenic activity that is powered by radioactive decay. The likelihood of internal differentiation, in turn, is directly related to global mass, and that can be deduced by a planetary body's influence on orbiting or passing probes and by the gravitational attraction it exerts on other planetary bodies or light.

To the extent that the evolution of life on Earth is a typical example, plate tectonics, which are driven by the internal heat of Earth, or some other effective recycling mechanism for minerals and nutrients, appears to be important for the persistence of living systems. Nutrients and minerals would otherwise be quickly exhausted and evolving life, especially when still in its infant stage and not well established, would not be able to meet its nutrient demands within a relatively short time frame (on a planetary time scale). On Earth and probably early Mars the recycling mechanism has been plate tectonics (Sleep 1994, Connerney et al. 1999). Plate tectonics on Earth have also constantly produced greenhouse gases that have acted as a global thermostat providing stability for the evolution of life (Ward and Brownlee 2000). The presence of plate tectonics can be identified with remote sensing methods based on measured magnetic properties of the rock, visible symmetry along a spreading axis, and specific patterns in fracture orientation and propagation.

9.2.3 Polymeric Chemistry

Chemical complexity is based at the molecular level on polymeric molecules joined by covalent bonds (Lwoff 1962). For reasons elaborated in Chap. 5, other life in the universe, except under very exotic conditions, is likely to be based on polymers of carbon. Polymeric organic compounds are in general detected by their absorption spectra.

On an active planet, polymeric organic compounds will be subject to chemical cycling. This can be inferred from spectra and gradients in surface coloration, and it appears to be widespread in our solar system. Io, Europa, Enceladus, Iapetus, and Triton, in addition to all the planets, provide examples. On Earth, chemical cycling occurs through oxidation-reduction reactions that are actively maintained by organisms, though they can occur inorganically as well.

9.2.4 Energy Source

A flow of energy is required to organize the material of the living state and to maintain its low entropic state (Morowitz 1968), thus an external energy source is a minimal requirement for life. Light and the oxidation of inorganic compounds provide the energy for the Earth's biosphere, so wherever light and a means for sustaining oxidation-reduction cycles can be demonstrated, the possibility for maintaining life is present. Light is a highly effective form of energy on Earth, and phototrophic organisms are responsible for the high oxygen content in the Earth's atmosphere. Light from the Sun could serve as the principal energy source for living systems on all the inner planets of our solar system, and possibly as far as the Jovian and Saturnian systems. Light is directly measurable using remote sensing and thus a good indicator for the theoretical possibility of photosynthesis. In general, energy gradients can fairly well be detected by remote sensing as detailed in Table 9.2. On Earth, all these energy sources are present. However, the availability of certain energy sources, such as heat, motion, or pressure, does not necessarily imply that life relies on them, but merely that the planetary body in question is active and meets one of the prerequisites for life.

Thermal gradients are commonly available energy sources throughout the inner solar system, and also among some larger satellites as shown from gradients of infrared and thermal infrared radiation (e.g. Io and Titan). Thermal energy can be derived from solar emissions or from radioactive heating if the mass of a planetary body is sufficient for differentiation into a radioactive core, as in the major Jovian satellites, and in Titan and Triton). Multispectral remote sensing methods are suitable for detecting rocks altered by hydrothermal heat and solutions, because their reflectance spectra differ from those of unaltered host rock. Thermal radiometry has been used, for instance, to determine that night-time temperatures on Europa are colder at the equator than at mid-latitudes for some longitudes, apparently due to latitude-dependent thermal inertia (Spencer et al. 1999 and 2001). Thermal radiometry has also been used extensively on Io (Spencer et al. 2000) and Mars (Christensen et al. 2003).

Kinetic energy is possible wherever gas or fluids exist. Atmospheric motion can be detected directly from visible clouds that move, such as those of Mars, Venus, the gas giant planets, and Titan, from visible and reflected infrared images. Speed and direction of moving objects can be determined by Doppler imaging at various wavelengths in an atmosphere or on a planetary surface. Active faults on which earthquakes may occur can be identified by observation of topographical features from space using radar (Tapponnier and Molnar 1977). Changes in the shape of a volcano caused by an expanding or contracting magma chamber can be determined by radar interferometry. Increased emissions of gas and heat of volcanoes can be identified with thermal infrared images and movements of plumes by images in the visible or infrared wavelengths. Electromagnetic measurements with the magnetometer instrument on board the Galileo orbiter were used to infer a conducting liquid in Europa's interior (Khurana et al. 1998, Kivelson et al. 2000).

Gravitational tides are exhibited by planetary bodies and major satellites that are in periodic alignments such as the Earth-Moon system, the Neptun-Triton system, and Jupiter and its four major moons. Significant tidal fluctuations in these sufficiently massive bodies are visible by the evidence of surface fragmentation and resurfacing of the planetary or lunar surface. For example, arcuate lineaments of vast extension on Europa have

Table 9.2 Remote detection of energy gradients (modified from Schulze-Makuch et al. 2002a).

Type of Energy Gradient	Examples within the solar system	Examples of Remote Detection
Light	Mercury to Saturnian system in sufficient intensity	Directly measurable
Chemical Cycling	Io, Europa, Iapetus, Triton	Molecular absorption spectra, surface reflectance spectra, imaging spectroscopy, polarimetry, radar measurements, detection of alteration minerals, gradients in surface coloration
Thermal	Mercury, Titan, Jovian satellites, Triton	Gradients of infrared radiation, thermal radiometry, infrared-to-visible spectral imaging, distance to Sun, mass sufficient for internal differentiation of planetary body (gravitational measurements), microwave radiometry to detect geothermal heat flows
Motion	Venus, Mars, Enceladus, Jovian satellites, Triton, Titan	Doppler imaging, radar interferometry, electromagnetic indications of a conducting liquid (e.g. Europa), thermal and infrared imaging (volcanic movement)
Gravitational Tides	Jovian satellites, Triton	Visible evidence of surface fragmentation and resurfacing, microwave radiometry
Pressure	Venus, Titan, Gas Giant planets	Visible clouds, changing atmospheric patterns (e.g. Red Spot on Jupiter)
Electromagnetism	Jovian and Saturnian system	Measurement of electromagnetic field fluctuations, detection of energetic particles

been interpreted as surface expressions of these enormous tidal forces (Hoppa et al. 1999).

Pressure gradients in an atmosphere can be inferred from banded cloud patterns and measurements of their rotational velocities by large storm systems such as Jupiter's famous Red Spot and less dramatic but similar examples on the other gas giants. Also, stratification as measured at Jupiter is another indicator of pressure gradients. Osmotic pressure gradients may also exist in planetary oceans that could be conducive for the support of life (Schulze-Makuch and Irwin 2002a). Those gradients are difficult to confirm directly, but could be inferred by a subsurface probe analyzing ocean chemistry, or possibly by remote determination of the solute content of a liquid eruption to the surface.

Electromagnetism is another energy source that occurs wherever electromagnetic fields are traversed or induced (Schulze-Makuch and Irwin 2001). Jupiter's magnetospheric plasma corotates with the planet at a velocity of 118 km/s, thereby creating a strong magnetic field (Beatty and Chaikin 1990). Energetic ion and electron intensities throughout the Jovian magnetosphere were measured by the Galileo orbiter using an energetic particle detector. Saturn generates a less massive but still large magnetosphere that will be mapped in detail by the Cassini orbiter. More benign electromagnetic fields and their fluctuations can be measured directly using a magnetometer.

9.2.5 Liquid Medium as a Solvent

Finally, a liquid medium appears to be favorable for living processes because macromolecules and nutrients can be concentrated within a bounded internal environment without immobilizing interacting constituents. This assumption is usually taken to mean an aqueous medium, though organic compounds and water mixtures with ammonia and other miscible molecules can exist in liquid form at temperatures well below the freezing point of water. The possibility that life could exist in dense atmospheres has also been suggested (Sagan and Salpeter 1976, Grinspoon 1997, Schulze-Makuch and Irwin 2002b). There is for example both experimental and observational evidence for organic synthesis in Jupiter's atmosphere (Sagan et al. 1967, Raulin and Bossard 1985, Guillemin 2000). It is difficult to envision how the boundary conditions necessary for compartmentalizing the flow of energy and restraining the population of interacting molecules could be established under such conditions. However, once originated in a liquid medium, life could adapt to thrive in a gaseous environment (Schulze-Makuch and Irwin 2002b, Schulze-Makuch et al. 2002b).

Liquid water is known for certain only on Earth, but may exist as subsurface water on Mars in underground aquifers (Carr 1986, Greeley 1987), or on Europa and Ganymede, where subsurface oceans are inferred from electromagnetic measurements from the Galileo orbiter (Khurana et al. 1998, Showman and Malhotra 1999) and from the presence of hydrated salt minerals on the surface (Kargel et al. 2000). Mixtures of water-ammonia-organic compounds are another possibility on cold planetary bodies, since these mixtures are liquid at much lower temperatures than water (Jakosky 1998). Theoretical models indicate the presence of subsurface stores that are liquid at extremely cold temperatures on Titan (Coustenis and Lorenz 1999, Fortes 2000) and possibly some of the satellites of Uranus and Neptune. Liquid water at or close to the surface can easily be detected by radar, gamma-ray spectrometry, and the absorption spectrum of water, but not when it is present in the deep subsurface or shielded by a thick layer of ice.

Liquid ethane and methane are assumed to be present on Titan's surface (Lorenz et al. 2003), and could provide an alternative solvent for life (Schulze-Makuch and Abbas 2004). Liquid sulfur compounds are inferred to exist on Io (Kieffer et al. 2000), and sulfur dioxide or hydrogen sulfide could play a role as solvent as well. Liquid compounds on a planetary surface can most easily be identified by visible and radar images of the erosional features that they cause.

9.3 Geoindicators for Life in our Solar System

Neither biosignatures nor geosignatures have been identified unambiguously on any planetary body beyond Earth to date, though the search should continue as resolution improves. Mars and Venus are the only other planetary bodies where life as we know it could plausibly be discovered by direct sampling in the foreseeable future. Thus, missions to Mars should remain a priority as they are currently with NASA and ESA, particularly lander missions (e.g. Viking 1 and 2, Pathfinder, Beagle 2, Spirit, Opportunity). In addition, the ease of reaching Venus and the possibility of an atmospheric habitat suitable for life there argue for an atmospheric sampling mission to Venus (Schulze-Makuch and Irwin 2002b, Schulze-Makuch et al. 2002b). In the meantime and for the coming decades, search for habitats suitable for life beyond the terrestrial inner planets of our solar system should focus on geoindicators such as those listed above. The current emphasis on visualization of surface features by the Mars Global Surveyor, the Mars Odyssey, and the recently terminated Galileo orbiter, and the visual data expected to be returned by the Cassini-Huygens Mission to Saturn and Titan are compatible with this strategy. These missions have the capability to detect energy gradients, organic chemicals, and near-subsurface as well as surface water.

The Huygens probe to Titan in January 2005, should add detailed knowledge of that body's atmosphere, weather, and surface chemistry. Because of the apparent similarity of its atmosphere to that of the early Earth, and its abundance of organic constituents, Titan should remain a high-priority target for exploration. In fact, geoindicators discussed here point to Titan as a suitable environment for life (Table 9.3), thus Titan should be considered a priority target of astrobiological significance. We previously proposed a plausibility of life index based on criteria for the existence of life such as the presence of (1) a fluid medium, (2) a source of energy, and (3) constituents and conditions compatible with polymeric chemistry under the key assumptions that (a) life arises quickly under appropriate formative conditions and (b) life remains static in stable environments or adapts to changing environments (Irwin and Schulze-Makuch 2001). We assigned a plausibility of life (POL) rating for each major planetary body in our solar system, while pointing out that the rating must be regarded as a dynamic value consistent with the information currently available. An updated definition of the plausibility of life (POL) categories with some examples are given in Table 9.4. An updated POL rating for all planets and major satellites in our solar system consistent with knowledge at the time of this writing is provided in Table 9.5.

Attention should also be given to some of the less known satellites. Organic constituents appear to be present on Triton and possibly Iapetus. Enceladus, Triton, and Titania show evidence of resurfacing that would indicate the generation of internal energy. Enceladus and Triton, along with Io, may be geothermally active. Higher resolution images and infrared data from Io would provide a more detailed picture of the complex thermal gradients on that body. Io may have also provided a suitable habitat for life early in the solar system, when it may have contained water on its surface. All these examples merit increased attention as future robotic exploration of the outer solar system is planned.

9.4 Extrasolar Planetary Detection

The possibility of worlds beyond our own has been appreciated since the speculations of the early Greeks. In 1584, Giordano Bruno asserted that there were "countless suns and countless earths all rotating around their suns", but confirmation of other solar systems with rotating, planar clouds of dust and gas that could lead to planet formation was not made until Bradford A. Smith, of the University of Arizona, and Richard J. Terrile, of the Jet Propulsion Laboratory, made infrared observations of such a disk of dust surrounding the star Beta Pictoris, in the 1980s (Smith and Terrile 1984).

The strategy for detecting planets around stars, based on perturbations in the star's motion (wobble) or appearance (from periodic occlusion) due to interaction with one or more sizeable, nearby planets has been pursued with vigor since the 1960s (Marcy and Butler 1998), but unambiguous evidence for an extrasolar planet did not come until the mid 1990s. Alexander Wolszczan (1994), a radio astronomer at Pennsylvania State University, discovered two or three planet-sized objects orbiting a pulsar in the Virgo constellation in 1994, and Michel Mayor and Didier Queloz in Geneva announced in 1995 that they had found a planet at least half the size of Jupiter rapidly orbiting the star 51 Pegasi (Mayor and Queloz 1995). This finding was soon confirmed by Geoff Marcy and his colleagues at the University of California, Berkeley (Marcy and Butler 1996, 1998), who have gone on to discover over 70 additional extrasolar planets. Since these initial findings, a flood of discoveries had brought the number of confirmed extra-solar planets to over 100 by the spring of 2003.

Because they are easier to detect, these planets have all been large like the gas giants of our outer solar system. Technologies currently operational or under development are hastening the discovery of planets such as these. The Large Binocular Telescope Interferometer is currently under construction in

Table 9.3 Presence of geoindicators for life on the planets and major satellites of our solar system based on current knowledge (modified from Schulze-Makuch et al. 2002a).

Major Planetary Body	Significant Atmosphere	Thermal Gradients and Chemical Disequilibrium	Internal Differentiation	Polymeric Chemistry	Energy Source	Liquid Solvent
Mercury	No	Yes	Yes	No	LH	None
Venus	Yes	Yes	Yes	Yes?	LCHP	H_2O, H_2SO_4?
Earth	Yes	Yes	Yes	Yes	LCHKGPM	H_2O
Moon	No	No	Yes	No	LG	None
Mars	Yes	Yes?	Yes	Unknown	LCH	H_2O
Jupiter	Yes	Yes	Yes	Yes	LCHKPM	?
Io	No	Yes?	Yes	Yes	CLHMG	H_2S?
Europa	Yes*	Unknown	Yes	Yes	CHKGOM	H_2O
Ganymede	Yes*	Unknown	Yes	Yes	CHKGOM	H_2O
Callisto	Yes*?	Unknown	Yes	Yes?	CHKGOM	H_2O
Saturn	Yes	Yes	Yes	Yes	CHKPM	None
Tethys	No	Unknown	No	Yes?	M	H_2O?
Dione	No	No	No	Yes?	M	H_2O?
Rhea	No	Unknown	No	Yes?	M	H_2O?
Enceladus	No	Yes?	No?	Yes	CHKM	H_2O
Iapetus	No	Yes?	No	Yes	CM	None?
Titan	Yes	Yes?	Yes	Yes	CHM	Ethane,CH_4?
Uranus	Yes	Yes?	Yes	Yes?	CHKPM	None
Titania	No	Yes?	Yes?	Yes?	CHG	H_2O
Ariel	No	No	No	No?	C?H?	None?
Miranda	No	No	No	No?	C?H?	None?
Umbriel	No	No	No	No?	C?H?	None?
Oberon	No	No	No	No?	C?H?	None?
Neptune	Yes	Yes?	Yes	Yes?	CHKPM	None?
Triton	Yes*	Unknown	Yes?	Yes	CHGO	H_2O/ NH_4/N_2?
Pluto/ Charon	No*?	Unknown	Yes?	Yes?	CG	H_2O/ NH_4/N_2?
Comets and Asteroids	No	No	Some	Some	L for some	None

Legend: L = light energy, C = chemical cycling, H = heat energy, K = kinetic energy (motion), G = gravitational energy (tides), P = pressure energy, O = osmotic gradients (in a possible high-salinity subsurface ocean), M = electromagnetic energy. Asterisks indicate a protective ice shield and trace atmosphere. Question marks indicate uncertainty, but with our estimate of probability in the indicated direction.

Table 9.4 Astrobiology plausibility categories (modifed from Irwin and Schulze-Makuch 2001).

Category	Defintion	Examples
I	Demonstrable presence of liquid water, readily available energy, and organic compounds	Earth
II	Evidence for the past or present existence of liquid water, availability of energy, and inference of organic compounds	Mars, Europa
III	Physically extreme conditions, but with evidence of energy sources and complex chemistry possibly suitable for life forms unknown on Earth	Venus, Titan, Triton, Enceladus
IV	Persistance of life very different from on Earth conceivable in isolated habitats or reasonable inference of past conditions suitable for the origin of life prior to the development of conditions so harsh as to make its perseverance at present unlikely but conceivable in isolated habitats	Mercury, Jupiter
V	Conditions so unfavorable for life by any reasonable definition that its origin or persistence cannot be rated a realistic probability	Sun, Moon

Arizona. It will combine the infrared light from two 8-meter class telescopes to provide interferometry capable of imaging distant galaxies and other faint objects over a wide field of view. The Keck Interferometer will combine light from the world's largest optical telescopes to enable the visualization of gas clouds, including large planets within them, around distant stars.

The proportion of planets that may be smaller and more terrestrial-like is unknown, but should be answered, at least for neighboring systems in our own galaxy, within the coming decade. The Space Interferometry Mission, scheduled for launch in 2009, will be able to measure distances with hundreds of times more precision than by present technologies, to the point of possibly detecting Earth-sized planets in other solar systems. In 2012, the Terrestrial Planet Finder is scheduled for launch. It will attempt to detect directly gases consistent with life such as carbon dioxide, ozone, and methane, of terrestrial-like atmospheres on planets around stars up to 45 light years away. The possibility of Earth-type habitable planets in other solar systems has been modeled and results indicate that this is a plausible scenario (e.g. planetary system of 47 UMa, Cuntz et al. 2003).

One of the most significant variables in the science of astrobiology is the unknown proportion of stars that contain planets, and the proportion of those that contain planets capable of harboring life. While it can no longer be doubted that planetary systems around other stars are relatively common, the frequency of terrestrial planets in those planetary systems remains unknown

Table 9.5 Plausibility of Life (POL) rating for planets and major moons in our solar system.

Body	POL Rating	Reasoning for Rating
Mercury	IV	Intense solar radiation; little if any geological cycling, water ice at poles, no atmosphere
Venus	II	Extreme heat at surface; highly caustic atmosphere; primordial ocean likely, water vapor in atmosphere, possibly supercritical water in the subsurface, minute amounts of organic compounds, extensive resurfacing implies geological activity and chemical recycling, possible life-supporting habitat in lower atmosphere
Earth	I	Life is present, salt water oceans and fresh water on surface, plate tectonics as geological recycling mechanism, ozone layer in oxygen-rich atmosphere
Mars	II	Oxidized surface, thin atmosphere, surface erosion by flowing water, not much geological cycling at present, liquid water on surface in the past; possible subsurface liquid water now; surface temperatures sometimes above the freezing point of water, polar ice caps with some water ice, presence of organic compounds inferred (e.g. by Martian meteorites)
Jupiter Saturn Uranus Neptune	IV IV IV IV	Gas giants with indistinct high-pressure atmosphere/ liquid transitions, presence of organic and nitrogen compounds, abundant energy but lack of solid substrates or sharp physical transitions; temperature and pressure extremes
Pluto	IV	Extreme cold, overall density ~ 2.1 implies composition of rock/ice mixture, mix of light and dark features implies complex chemistry, possible subsurface liquids due to tidal interaction with Charon
Moon	V	Extremely dry, no protective atmosphere, water ice at poles, no geological cycling
Io	III	Sharp thermal gradients and geochemical cycling; but harsh temperature and radiation, volcanic activity generates thin atmosphere and liquid sulfur compounds near surface for some time periods, surface coloration implies complex chemistry
Europa Ganymede	II II	Planetary ocean likely beneath ice shell, surface coloration implies complex chemistry and chemical cycling, high radiation doses and low temperature at water ice surface
Callisto	III	Possible subsurface liquid water, but little energy flux, low density implies mostly water-ice, high radiation doses and low temperature at ice surface
Tethys Dione Rhea	IV IV IV	Little evidence for liquid water at present, very low density and high albedo implies mostly water-ice composition, high radiation environment, very cold

Table 9.5 (continued)

Body	POL Rating	Reasoning for Rating
Enceladus	III	Extensive resurfacing, with evidence of ice geysers, possible subsurface liquid water, high radiation environment, very cold, too small for significant core and radiogenic heating to be present
Iapetus	IV	No evidence for liquid water at present, low density and moderate albedo implies mostly ice composition, dark leading edge implies possible hydrocarbon chemistry, high radiation environment, very cold
Titan	II	Complex organic chemistry and reducing atmosphere, hydrocarbon liquids present on surface, density of about 1.8 implies organic liquids and/or water-ice with solid core, dense, colored atmosphere implies complex chemistry
Titania	III	Possible subsurface or recent surface liquid with evidence of liquid flow in canyons, relatively small for radiogenic heating to occur, very cold at surface
Ariel Miranda Umbriel Oberon	IV IV IV IV	Small size and insufficient evidence for energy gradients, high albedo and density of about 1.5–1.7 implies rock/ice composition, very cold.
Triton	III	Complex chemistry and several energy sources, with possible subsurface liquid water, density of about 2 implies rocky core with water/ice surface, surface coloration implies complex chemistry, unusual surface features imply internal energy, elliptical orbit implies tidal flexing and temperatures varying with seasons
Charon	IV	Extreme cold, density of about 2.1 implies rock/ice mixture, mix of light and dark features implies complex chemistry
Comets and Asteroids	V	Extreme cold, no atmosphere, no persistent internal energy source, rock/ice mixtures in composition, abundant water ice with possible hydrothermal alteration in parent bodies

at the present time. If we assume that our solar system is not extraordinary, the frequency will likely be high, with a correspondingly high probability that life as we know it could have evolved and may be flourishing throughout the universe. Until that assumption is confirmed, however, the more conservative view that terrestrial planets are rare (Ward and Brownlee 2000) must be considered a serious possibility. In the latter case, the chances for life as we know it are diminished in frequency, but not rendered implausible. Unlike a number of areas of astrobiology, this is a question that data acquired within the coming decade will probably be able to answer.

Much of the search for extrasolar planets is motivated by the quest for terrestrial planets, for the common-sense reason that we are better qualified to recognize life as we know it, and therefore more likely to find it on smaller, rocky planets such as our own. It is important to consider, however – as we argue at numerous points in this book – that life in forms unfamiliar to us could flourish under conditions alien to the life with which we are familiar. There may thus be specialized niches for some forms of life on gas giants, or on their satellites, or on brown dwarfs or orphan planets, within radiation fields of high intensity, in liquids other than water, using metabolic systems and energy sources unlike anything we have ever seen. We have already been surprised to find, for example, that our closest star belongs to a triple star system (Alpha Centauri A, Alpha Centauri B, and Proxima Centauri), and we have just started to explore other solar systems. The broader mandate for space and planetary science should therefore be to characterize the full range and variety of solar systems, and seek in the pattern of their distribution the clues that will lead us to consider how exotic our consideration of life on other worlds should remain.

9.5 Chapter Summary

The search for extraterrestrial life everywhere but on our closest planetary neighbors (Mars and Venus) is limited for the foreseeable future by our inability to obtain physical samples. Therefore, information that can only be obtained by remote sensing and robotic probes will for now provide the only clues concerning the existence of life elsewhere. The search parameters we have proposed emphasize the importance of detecting the presence of physical and chemical gradients of all kinds, because of their potential for generating free energy. Other geoindicators that would enhance the prospects for life include evidence for polymeric chemistry in association with chemical cycling, the presence of an atmosphere or ice shield, sufficient mass for endogenic heating, and the availability of a liquid that may act as a solvent to enhance chemical reactions. Also, any unusual topographical features or surface patterns that cannot be easily explained by well-understood geological and geochemical processes should be regarded as evidence for the possibility of environmental changes induced by living systems.

References

Akasofu S (1999) Auroral spectra as a tool for detecting extraterrestrial life. *EOS Transactions AGU* 80: 397.

Beatty JK, Chaikin A (1990) *The new solar system.* 3rd edition, Sky Publishing Corporation, Cambridge, Massachusetts.

Boston PJ, Ivanov MV, McKay CP (1992) On the possibility of chemosynthetic ecosystems in subsurface habitats on Mars. *Icarus* 95: 300–308.

Carr MH (1986) Mars: A water rich planet. *Icarus* 56: 187–216.

Christensen PR, Bandfield JL, Bell JF, Gorelick N, Hamilton VE, Ivanov A, Jakosky BM, Kieffer HH, Lane MD, Jakosky BM, Kieffer HH, Lane MD, Malin MC, McConnochie T, McEwen AS, McSween HY, Mehall GL, Moersch JE, Nealson KH, Rice JW, Richardson MI, Ruff SW, Smith MD, Titus TN, Wyatt MB (2003) Morphology and composition of the surface of Mars: Mars Odyssey THEMIS results. *Science* 300: 2056–2061.

Connerney JEP, Acuna MH, Wasilewski PJ, Ness NF, Reme H, Mazelle C, Vignes D, Lin RP, Mitchell DL, Cloutier PA (1999) Magnetic lineations in the ancient crust of Mars. *Science* 284: 794–798.

Coustenis A, Lorenz RD (1999) Titan. In: Weissman PR, Johnson ATV (eds) *Encyclopedia of the Solar System.* Academic Press, New York, pp 377–404.

Cuntz M, Von Bloh W, Bounama C, Franck S (2003) On the possibility of Earth-type habitable planets around 47 Uma. *Icarus* 162: 214–221.

Diaspro A, Bertolotto M, Vergani L, Nicolini C (1991) Polarized light scattering of nucleosomes and polynucleosomes – in situ and in vitro studies, *IEEE Trans. Biomed. Eng.* 38: 670–678.

Dyson FJ (1959) Search for artificial sources of infrared radiation. *Science* 131: 1667.

Fortes AD (2000) Exobiological implications of a possible ammonia-water ocean inside Titan. *Icarus* 146: 444–452.

Frey HU, Lummerzheim D (2002) Can conditions for life be inferred from optical emissions of extra-solar-system planets. In *Atmospheres in the Solar System: Comparative Aeronomy. Geophysical Monograph* 130, American Geophysical Union, pp 381–388.

Gervin JC, Kerber AG, Witt RG, Lu YC, Sekhon R (1985) Comparison of level I land cover accuracy for MSS and AVHRR data. *International Journal of Remote Sensing* 6: 47–57.

Gorbushina AA, Krumbein WE, Volkmann M (2002) Rock surfaces as life indicators: new ways to demonstrate life and traces of former life. *Astrobiology* 2: 203–213.

Gordon HR, Clark DK, Mueller JL, Hovis WA (1980) Phytoplankton pigments from the Nimbus-7 coastal zone color scanner – comparison with surface measurements. *Science* 210: 63–66.

Greeley R (1987) Release of juvenile water on Mars – estimated amounts and timing associated with volcanism. *Science* 236: 1653–1654.

Griffith CA, Owen T, Wagener R (1991) Titan's surface and tropossphere, investigated with ground-based near-infrared observations. *Icarus* 93: 362–378.

Grinspoon DH (1997) *Venus revealed: a new look below the clouds of our mysterious twin planet.* Perseus Publishing, Cambridge, Massachusetts.

Guillemin J-C (2000) Organic photochemistry in the atmosphere of Jupiter and Saturn – The role played by H_2S, PH_3 and NH_3. *Origins of Life Evol. Biosphere* 30: 236.

Hall DT, Strobel DF, Feldman PD, McGrath MA, Weaver HA (1995) Detection of an oxygen atmosphere on Jupiter's moon Europa. *Nature* 373: 677–679.

Hoppa GV, Tufts BR, Greenberg R, Geissler PE (1999) Formation of cycloidal features on Europa. *Science* 285: 1899–1902.

Hovis WA, Clark DK, Anderson F, Austin RW, Wilson WH, Baker ET, Ball D, Gordon HR, Mueller JL, Elsayed SZ, Sturm B, Wrigley RC, Yentsch CS (1980) Nimbus-7 CZCS coastal zone color scanner – system description and early imagery. *Science* 210: 60–63.

Irwin LN, Schulze-Makuch D (2001) Assessing the plausibility of life on other worlds. *Astrobiology* 1: 143–160.

Jakosky B (1998) *The Search for Life on Other Planets.* Cambridge University Press, Cambridge, UK.

Kargel JS, Kaye JZ, Head III JW, Marion GM, Sassen R, Crowley JK, Ballesteros OP, Grant SA, Hogenboom DL (2000) Europa's crust and ocean: origin, composition, and the prospects for life. *Icarus* 148: 226–265

Khurana KK, Kivelson MG, Stevenson DJ, Schubert G, Russell CT, Walker RJ, Polanskey C (1998) Induced magnetic fields as evidence for subsurface oceans in Europa and Callistro. *Nature* 395: 777–780.

Kieffer SW, Lopes-Gautier R, McEwen A, Smythe W, Keszthely L, Carlson R (2000) Prometheus: Io's wandering plume. *Science* 288: 1204–1208.

Kivelson MG, Khurana KK, Russell CT, Volwerk M, Walker RJ, Zimmer C (2000) Galileo magnetometer measurements: a stronger case for a subsurface ocean at Europa. *Science* 289: 1340–1343.

Leger A, Pirre M, Marceau FJ (1993) Search for primitive life on a distant planet: relevance of O_2 and O_3 detections. *Astronomy and Astrophysics* 277: 309–313.

Lofftus KD, Quinby-Hunt MS, Hunt AJ, Livolant F, Maestre M (1992) Light scattering by Prorocentrum micans: a new method and results, *Applied Optics* 31: 2924–2931.

Lorenz RD (1993) The surface of Titan in the context of the ESA Huygens probe. *ESA J.* 17: 275–292.

Lorenz RD, Lunine JI (1997) Titan's surface reviewed: the nature of bright and dark terrain. *Planet. Space Sci.* 45: 981–992.

Lorenz RD, Kraal E, Asphaug E, Thomson RE (2003) The seas of Titan. *EOS Transactions AGU* 84: 125 & 131–132.

Lovejoy JE (2000) The Gaia hypothesis. In: Margulis L, Matthews C, Haselton A (eds) *Environmental Evolution.* MIT Press, Cambridge, Massachusetts, pp 1–28.

Lunine JI, Yung YL, Lorenz RD (1999) On the volatile inventory of Titan from isotopic abundances in nitrogen and methane. *Planetary Space and Science* 47: 1291–1303.
Lwoff A (1962) *Biological order.* MIT Press, Cambridge, Massachusetts.
Marcy GW, Butler RP (1996) A planetary companion to 70 Virginis. *Astrophysical Journal* 464: L147–L151.
Marcy GW, Butler RP (1998) Detection of extrasolar giant planets. *Annual Review of Astronomy and Astrophysics* 36: 57–97.
Margulis L (1998) *Symbiotic planet.* Sciencewriters, Brockman, Inc., Amherst, Massachusetts.
Mayor M, Queloz D (1995) A Jupiter-mass companion to a solar-type star. *Nature* 378: 355–359.
McKay DS, Everett KG, Thomas-Keprta KL, Vali H, Romanek CS, Clemett SJ, Chillier XDF, Maechling CR, Zare RN (1996) Search for past life on Mars: possible relic biogenic activity in Martian Meteorite ALH84001. *Science* 273: 924–930.
Morowitz HJ (1968) *Energy flow in biology.* Academic Press, New York.
Nicolini C, Diaspro A, Bertolotto M, Facci P, Vergani L (1991) Changes in DNA superhelical density monitored by polarized light scattering. *Biochemical and Biophysical Research Communications* 177: 1313–1318.
Pasteris JD, Wopenka B, Schopf JW, Kudryavtsev AA, David G, Wdowiak TJ, Czaja AD (2002) Images of Earth's earliest fossils?; discussion and reply. *Nature* 420: 476–477.
Plaxco KW, Allen SJ (2002) Life detection via tetrahertz circular dichroism spectroscopy. *American Geophysical Union Fall Meeting Abstracts* U61B–02 p F6.
Raulin F, Bossard A (1985) Organic synthesis in gas phase and chemical evolution in planetary atmospheres. *Adv. Space Res.* 4: 75–82.
Sagan C (1994) *Pale Blue Dot.* New York: Random House, New York.
Sagan C, Lippincott ER, Dayhoff MO, Eck RV (1967) Organic molecules and the coloration of Jupiter. *Nature* 213: 273–274.
Sagan C, Salpeter EE (1976) Particles, environments, and possible ecologies in the jovian atmosphere. *Astrophys. J. Suppl. Ser.* 32: 624.
Salzman GC, Griffith JK, Gregg CT (1982) Rapid identification of microorganisms by circular-intensity differential scattering. *Applied and Environmental Microbiology* 44: 1081–1085.
Schidlowski M (1988) A 3,800-million-year isotopic record of life from carbon in sedimentary rocks. *Nature* 333: 313–318.
Schidlowski M, Hayes JM, Kaplan IR (1983) Isotopic inferences of ancient biochemistry: carbon, sulfur, hydrogen, and nitrogen. In: Schopf JW (ed) *Earth's earliest biosphere: its origin and evolution.* University Press, Princeton, New Jersey, pp 149–186.
Schroedinger E (1944) *What is life?: the physical aspect of the living cell.* Cambridge University Press, Cambridge, UK.

Schulze-Makuch D (2002) At the crossroads between microbiology and planetology: a proposed iron redox cycle could sustain life in an ocean – the the ocean need not be on Earth. *ASM News* 68: 364–365.

Schulze-Makuch D (2004) Possible microbial habitats and metabolisms on Titan. Presentation given at "The Limits of Organic Life in Planetary Systems", National Academy of Sciences, USA, 10–12 May 2004, Washington, D.C.

Schulze-Makuch D, Abbas O (2004) Titan: a prime example for alternative possibilities of life? In: Simakov M (ed) *Exobiology of Titan*, in press.

Schulze-Makuch D, Irwin LN (2001) Alternative energy sources could support life on Europa. *EOS, Transactions American Geophysical Union* 82: p 150.

Schulze-Makuch D, Irwin LN (2002a) Energy cycling and hypothetical organisms in Europa's ocean. *Astrobiology* 2: 105–121.

Schulze-Makuch D, Irwin LN (2002b) Reassessing the possibility of life on Venus: proposal for an astrobiology mission. *Astrobiology* 2: 197–202.

Schulze-Makuch D, Irwin LN, Guan H (2002a) Search parameters for the remote detection of extraterrestrial life. *Planet. Space Sci.* 50: 675–683.

Schulze-Makuch D, Irwin LN, Irwin T (2002b) Astrobiological relevance and feasibility of a sample collection mission to the atmosphere of Venus. Proceedings of the 2nd European Workshop on Exo-Astrobiology (EANA/ESA), Graz, Austria, 16–19 Sept. 2002, ESA SP–518: 247–250.

Showman AP, Malhotra R (1999) The Galilean satellites. *Science* 286: 77–84.

Sleep NH (1994) Martian plate tectonics. *J. Geophys. Res.* 99: 5639.

Smith BA, Terrile RJ (1984) A circumstellar disk around Beta-Pictoris. *Science* 226: 1421–1424.

Spencer JR, Tamppari LK, Martin TZ, Travis LD (1999) Temperatures on Europa from Galileo photopolarimeter-radiometer. *Science* 284: 1514–1516.

Spencer JR, Rathbun JA, Travis LD, Tamppari LK, Barnard L, Martin TZ, McEwen AS (2000) Io's thermal emission from the Galileo photopolarimeter-radiometer. *Science* 288: 1198–1201.

Spencer JR, Martin TZ, Goguen J, Tamppari LK, Barnard L, Travis LD (2001) Galileo PPR observations of the Galilean satellites. In *Jupiter: Planets, Satellites & Magnetosphere*, Abstract volume, Boulder Colorado, pp 105–106.

Tapponnier P, Molnar P (1977) Active faulting and tectonics in China. *Journal of Geophysical Research* 82: 2905–2930.

Tucker CJ, Townshend JRG, Goff TE (1985) African land-cover classification using satellite data. *Science* 227: 369–375.

Van Holde KE, Johnson WC, Ho PS (1998) *Principles of physical biochemistry*, Prentice Hall, Upper Saddle River, New Jeryes.

Ward PD, Brownlee D (2000) *Rare Earth: why complex life is uncommon in the universe.* Copernicus, New York.

Wolszczan A (1994) Confirmation of Earth-mass planets orbiting the millisecond pulsar Psr B1257+12. *Science* 264: 538–542.
Xu J, Ramian GJ, Galan JF, Savvidis PG, Scopatz AM, Birge RR, Allen SJ, Plaxco KW (2003) Terahertz circular dichroism spectroscopy: a potential approach to unbiased, in situ life detection. *Astrobiology* 3: 489–504.

Printing: Mercedes-Druck, Berlin
Binding: Stein + Lehmann, Berlin